OXFORD
INDIA SHORT
INTRODUCTIONS

ARTIFICIAL
INTELLIGENCE AND
INDIA

The Oxford India Short
Introductions are concise,
stimulating, and accessible guides
to different aspects of India.
Combining authoritative analysis,
new ideas, and diverse perspectives,
they discuss subjects which are
topical yet enduring, as also
emerging areas of study and debate.

SOME OTHER TITLES IN THE SERIES

Kashmir
Chitralekha Zutshi

Employment in India
Ajit K. Ghose

Indian Federalism
Louise Tillin

Surrogacy
Anindita Majumdar

Jawaharlal Nehru
Rudrangshu Mukherjee

The Partition of India
Haimanti Roy

Indian Nuclear Policy
Harsh V. Pant and Yogesh Joshi

Indian Democracy
Suhas Palshikar

Indian National Security
Chris Ogden

Indian Foreign Policy
Sumit Ganguly

Dalit Assertion
Sudha Pai

For more information, visit our website:
https://india.oup.com/content/series/o/
oxford–india–short–introductions/

OXFORD
INDIA SHORT
INTRODUCTIONS

ARTIFICIAL
INTELLIGENCE
AND INDIA

KAUSHIKI SANYAL
RAJESH CHAKRABARTI

OXFORD
UNIVERSITY PRESS

OXFORD
UNIVERSITY PRESS

Oxford University Press is a department of the University of Oxford.
It furthers the University's objective of excellence in research, scholarship,
and education by publishing worldwide. Oxford is a registered trademark of
Oxford University Press in the UK and in certain other countries.

Published in India by
Oxford University Press
22 Workspace, 2nd Floor, 1/22 Asaf Ali Road, New Delhi 110 002

ISBN-13 (print edition): 978-0-19-012854-8
ISBN-10 (print edition): 0-19-012854-2

ISBN-13 (eBook): 978-0-19-099205-7
ISBN-10 (eBook): 0-19-099205-0

Typeset in 11/14.3 Bembo Std
by The Graphics Solution, New Delhi 110 092
Printed in India by Rakmo Press, New Delhi 110 020

To Kajori

Contents

1
The Big Picture

Science Fiction No Longer, and Not Foreign Either

On a bright Sunday afternoon in Guangzhou, China, 23-year-old Changpu Chen nervously tightened his tie-knot in his student housing room before starting his interview with a mid-level real estate firm. The job he had applied for would involve selling to international clients, so Changpu knew his spoken English would be an important factor. He relaxed gradually as the interview progressed and focused his mind on the questions.

Little did Changpu realize that at the other end of his connection was a computer system evaluating him using a software designed in India. In addition to evaluating his responses and feeding him questions accordingly, it also recorded non-verbal cues from his body language, facial expressions, and voice modulations that went into an algorithm honed through months of machine learning.

The creator of the software was a start-up founded by the Aggarwal brothers Himanshu and Varun, with Professor Tarun Khanna, faculty at the Harvard Business School (HBS). In 2007, Himanshu, an IIT Delhi alumnus, was looking for his next venture after founding an online start-up in India. He had also completed a stint at the California-based company NetApp. Varun had completed his master's in Electrical Engineering and Computer Science from MIT. The firm Himanshu and Varun set up, Aspiring Minds, focused on the area of candidate assessment and skill certification, a crying need in a country producing annually as many engineers as the population of Australia, with less than a quarter of them employable. The years since have been a heady ride for Aspiring Minds and its journey mirrors, to a significant extent, the adoption and development of artificial intelligence in India.

The first product of the Aggarwal brothers' company—the Aspiring Minds Computer Adaptive Test (AMCAT)—was designed for Information Technology (IT) employers who were mostly looking for analytical skills and domain knowledge together with functional English language proficiency and certain personality traits in their employees. The product hit the market during the global recession of 2008, with tech recruitment coming almost to a grinding halt. It took the firm a year to pivot to the sectors that were still hiring—banking and microfinance. But as it sought out new clients across industries, the firm had to choose between the demand from HR managers for customized tests picking up sector-specific competencies and continuing with its standardized test. The compromise was a standardized test with component-wise results.

The choice mattered, for at the heart of the AMCAT lay a data-driven learning strategy. For every company that signed on, the test could be conducted on its existing employees and the results co-related to their performance data. AMCAT was learning what really mattered in the workplace and sharpening itself continuously. In parallel, the product was being sold to job-seekers at universities. The AMCAT score could be a valuable addition to the typical sparse resume of a fresher in the job market. But the core competitive advantage of Aspiring Minds lay in its machine learning competence—to be able to create tests that could predict workplace performance in a wide range of industries. Before 2010, machine learning was still a fairly esoteric concept in India, almost completely restricted to elite engineering departments. AMCAT remained hungry for data and kept on adapting itself as its reach grew among Indian corporates.

The research team at Aspiring Minds, which expanded to include 15 hand-picked people in the first 8 years, was a key differentiator. The unwritten membership rule was simple: one had to be good enough to be recruited by a computer science graduate programme in one of the top 15 computer science departments in the USA. Varun stayed close to academia, presenting at conferences, and the team published its research on machine learning in peer-reviewed journals. Developing AMCAT—marrying cutting-edge machine learning with the psychology of education—was back-breaking research work, but what made it a real success was its ability to learn continuously. For every test-taker that used AMCAT, not only was the product providing value to users, but it was getting another data point to improve itself.

In 2011 came Aspiring Minds' next product, SVAR, focused on English speaking skills. It was a phone-based test where the test-taker called a number and responded to questions. Pronunciation, fluency, listening skills, and grammar were among the competencies that were tested.

Beyond the sheer technological challenge, the cost advantage of being based in India kicked in in a major way. SVAR was packaged with AMCAT and the latter's USD 15 fee covered its cost. Pearson, the main rival in this space, had a computer-based product that cost the user USD 200.

In 2013 came Automata, a test to capture programming skills, followed by Customer Service for the service industry, and the TESLA suite of products for vocational skills such as plumbing, carpentry, and construction. As the skill development sector exploded, TESLA was adopted in 20 to 30 Sector Skill Councils set up by the National Skill Development Corporation (NSDC) to assess and certify workers. By 2015 Aspiring Minds was earning a third of its revenue from this area with no direct competitor in vocational assessment.

Among the latest offerings of Aspiring Minds is AutoView, an automated AI-based evaluation tool on a virtual interview platform that can mimic face-to-face interviews. It observes candidates on most dimensions that human interviewers can consciously identify including emotional intelligence captured through body language, facial expressions, and voice modulations.

Success came relatively early to Aspiring Minds. It saw positive cash flows within three years of being founded and by the next five had established itself as the industry standard in the country in almost every segment it entered. Now the wider world beckoned. The India advantage worked here as well. A product honed in the diversity and complexity of the Indian context was likely to do well in most other settings though the competencies might be considerably different. Starting with Tanzania in 2015, Aspiring Minds expanded rapidly to the Middle East and Philippines before establishing its US headquarters in Redwood, California, in a single year.

Aspiring Minds entered China in the same year. Here was a new world of opportunity for it. The market for assessments in China was about five times bigger than India with a greater willingness to pay. But there were challenges as well. The bureaucracy of permits was a pain, even for Indian founders. Operationally, language created its own set of challenges to product customization, though they appeared to be short-term ones. But a new set of challenges appeared through the question of data location. Servers in India and the USA could serve all geographies for Aspiring Minds, but not China. Also, culture and language issues reduced the effectiveness of the 'home team' in China. A local Chinese team became inevitable. With China presenting such vast opportunities and challenges, Himanshu had to quickly shift base there to handle matters on the spot.

The Aspiring Minds success story gives a glimpse into how India has started to leverage its advantage—demographic and technological—as the world grapples with a new disruptive technology: Artificial Intelligence

(AI). It also shows the potential for machine learning and automation across many sectors and roles. Is the Aspiring Minds story typical of most AI start-ups in India? If not, what is the real picture? How prepared is India to not only cope with this technology but take advantage of its potential? Where does India stand globally in terms of research, investment, and trained manpower for AI? What are the potential challenges and ethical issues that will have to be tackled?

What to Expect from This Book

For many, the term 'AI' still evokes the image of either a Terminator-type robot or a disembodied talking computer many times smarter than humans. And more often than not, the human race invariably needs to be rescued from their clutches preferably by a bunch of easy on the eyes Hollywood stars. Fortunately, the reality is considerably less dramatic and more benign. A conscious computer or super intelligent robot is what experts refer to as Artificial General Intelligence or Singularity and we are still about 40 to 50 years away from the kind of robots mentioned in Isaac Asimov's books.

However, narrow AI is already in our midst—in our smartphones talking to us (Siri, Cortana), giving us recommendations (Amazon and Netflix), giving financial advice (Schwab's Intelligent Portfolio), and winning game shows (IBM's Watson). The common thread among all these different activities performed by non-humans is the fact that they are replicating what a reasonably smart human being can do—sensing, reasoning, and acting. Put simply, AI is the ability of machines to perform functions similar to that of a human mind. It is a sub-field of computer science and is aimed at developing a set of computational technologies that are capable of doing things that are traditionally done by people.

Currently, there is considerable confusion about the boundaries of AI among the lay people. Technologies such as blockchain, Internet of Things (IoT), and big data are often used interchangeably with AI. However, while they can all be classified as emerging technologies, that is where the similarity ends. Each of these technologies have specific characteristics and applications, and can be combined with AI for certain functionalities. But, strictly speaking, in and of themselves, they do not constitute AI. Chapter 2 delineates the contours of AI and provides a sneak peek into its evolution, globally and in India.

AI is also a catch-all term for an array of disciplines and segments. Subsections of AI include robotics used mostly in factories and assembly lines; computer vision used in applications from identifying a photo to self-driving car technology; Natural Language Processing used mostly in translation; speech recognition used in voice assistants on phones and home speakers; and advanced analytics represented by machine learning algorithms that work on large data sets to make predictions, for forecasting, and for risk analysis. What is important to understand is that there is virtually no limit to the forms AI can or will take now or in the future in interacting with us.

How did the discipline of AI evolve into its current form from its generally accepted inception about 65 years ago at a Dartmouth Conference?[1] The short answer is: in fits and starts. But the last two decades have changed that as progress in the field has accelerated. Once an area dominated by nerds, AI has enabled a constellation of mainstream technologies in the twenty-first century that are having a substantial impact on our everyday lives. AI is being deployed in diverse sectors such as finance, national security, health care, criminal justice, transportation, and smart cities. There are numerous examples where AI is already making an impact on the world and augmenting human capabilities in significant ways. One of the

[1] The Dartmouth Conference was held in 1955, in which John McCarthy, widely hailed as the father of AI, defined the core mission of AI.

reasons for the growing role of AI is the tremendous opportunities for economic development that it presents. PriceWaterhouseCoopers (PwC) estimated that 'artificial intelligence technologies could increase global GDP by USD 15.7 trillion, a full 14 per cent, by 2030'.[2] That includes advances of USD 7 trillion in China, USD 3.7 trillion in North America, USD 1.8 trillion in Northern Europe, USD 1.2 trillion in Africa and Oceania, USD 0.9 trillion in the rest of Asia outside of China, USD 0.7 trillion in Southern Europe, and USD 0.5 trillion in Latin America. Other studies have made similar conclusions about the economic impact of AI.

However, the impact and application of AI is not limited to increasing efficiency and productivity in businesses or development of new products. AI is also addressing many of society's most pressing issues from climate change to better policing and access to quality health care and education. For instance, AI has been applied to detect congenital heart disorders, cancer, and diabetes. AI also has applications in the military and in

---

[2] PriceWaterhouseCoopers, 'Sizing the Prize: What's the Real Value of AI for Your Business and How Can You Capitalise?' 2017, available at https://www.pwc.com/gx/en/issues/analytics/assets/pwc-ai-analysis-sizing-the-prize-report.pdf, last accessed on 18 February 2020.

the field of law enforcement. The city of Chicago has developed an AI-driven 'Strategic Subject List'. This system analyses the previous arrest record of people to predict their risk of becoming future perpetrators. It ranks 400,000 people on a scale of 0 to 500 using items such as age, criminal activity, victimization, drug arrest records, and gang affiliation. Through its 'Sharp Eyes' programme, Chinese law enforcement agencies are storing video images, social media activity, records of online purchases, travel records, and personal identity of individuals on a 'police cloud'. This integrated database enables authorities to keep track of criminals, potential lawbreakers, and terrorists.

Where does India stand in this AI revolution sweeping through the world? Is our workforce prepared for the new forms of technology? How well-equipped are our industries to adopt AI-driven technologies? What role is our government playing in facilitating and regulating AI?

India's Position in the AI Ecosystem

Economist Joan Robinson once famously observed, 'Whatever you can rightly say about India, the opposite is also true.' This applies to AI with a vengeance. India is among the top 10 countries in terms of being home to AI start-ups but it lags far behind in private sector investment in AI. According to a report by LinkedIn, India's workforce is equipped with AI skills but most of its engineering graduates are unemployable. India ranks third in AI research but it only has 50 to 75 principal researchers in the country and internet penetration is as low as 32 per cent.[3]

Shortage of data and lack of data sharing protocols, lack of AI level programming abilities and imagination, and higher cost of AI implementation have all combined to delay the march of AI and related technology in India. Only the fintech sector appears to have been somewhat of an exception to that rule.

That, however, appears to be a story of the past. In the past two to three years, AI and related technologies have started receiving adequate attention in both private and public sectors in India, and applications by MNCs, Indian start-ups, large firms, and the government alike are finding rapid acceptance. Nevertheless, in both public discourse as well as business and policy circles, AI remains a 'curiosity item' rather than a regular occurrence.

Although India is way behind the USA and China in terms of investment, it is among the top players

---

[3] A principal researcher is responsible for the research design, the conduct of research, supervision of any research staff, and the research findings.

when it comes to generation of skilled manpower and its companies are among early adopters of AI technology. NASSCOM's data shows that in 2017 investors have started focusing on more mature start-ups with increased investment but fewer new start-ups are being incorporated. Top recent investments ranging from USD 19 million to USD 1 million were bagged by SigTuple, Active Intelligence, Observe, Uniphore, CreditVidya, and Edge Network. Companies are building AI platforms to industrialize their offerings. Some examples include Wipro Homes, HCL DRYiCE, IBM Watson, and Capgemini InSTREAM. Some of the key investors in AI are ACCEL Partners, IDG Ventures India, Endiya, Sequoia, Matrix Partners, Y Combinator, Axilor, and Blume Ventures.

In addition, there are a number of incubators and accelerators to provide support to start-ups, such as Avishkar, Pitney Bowes Accelerator Program, Jio Gennext, The Hive, and Z Nation Labs.

NASSCOM predicts that AI has the potential to add USD 957 billion to the Indian economy by 2035 if there is a focused push across a few sectors. The government is also taking an active interest in promoting AI through schemes such as Start-up India, Make-in-India, and most recently the National Mission on AI.

Chapter 3 looks at ways in which AI is beginning to transform Indian industry.

AI in Public Policy

In addition to fast-adopting sectors such as IT, fintech, retail, and defence, in India sectors such as education, agriculture, law enforcement, and public service delivery are also adopting AI.

Consider the area of law enforcement, for instance. While still in the early stages of developing the technological proficiency to fully implement AI solutions in law enforcement, India has made some strides towards the use of big data analytics and algorithms for the purpose of processing vast tracts of data to generate predictive policing models, and states such as Kerala, Odisha, and Maharashtra may adopt them soon. In 2017 law enforcement agencies in Rajasthan commissioned a pilot project with Staqu, an AI start-up, to develop the application ABHED (Artificial intelligence Based Human Efface Detection) for facilitating criminal identity registration, tracking, and missing persons search. A Hyderabad-based technology start-up, H-Bots Robotics has developed a smart policing robot that has yet to be deployed in the field. The 'robocop' can play a role in handling law and order, and in enhancing traffic management.

In the education sector, AI is predominately being used in decision making, student services, student progress monitoring, and personalized learning.

US-based service provider HTC Global Services is focusing on launching a product in the education space in India which is a web-based application that will enable students make more well thought out decisions when choosing courses and electives at universities. The Chandrababu Naidu-led government in Andhra Pradesh is looking to collect information from a variety of databases and process the data through Microsoft's Machine Learning Platform to enable personalized monitoring of children and devote individualized attention to their progress, curbing school dropout rates. Ek-step is an open-learning platform that uses Application Programming Interfaces (API) that uses gamified apps available on Google Play Store. It is being used in government schools across Karnataka and there are plans to launch it across the country.

In agriculture, the principal application of AI in India has been in the domain of predictive analytics. Microsoft collaborated with the International Crops Research Institute for the Semi-Arid Tropics (ICRISAT) to develop an AI Sowing App powered by the Microsoft Cortana Intelligence Suite, which sends advisories to farmers providing them with information on the optimal date to sow.

Public service delivery, a massive policy challenge for India, is beginning to embrace AI. The government has started leveraging AI to deliver services to the citizens.

The Andhra Pradesh government has partnered with Microsoft to develop the Kaizala app to crowdsource citizen grievances. AI is also being used for monitoring the implementation of governance projects. For example, the National Informatics Centre is piloting a project that will monitor a toilet construction programme under the Swachh Bharat Abhiyan.

Chapter 4 looks at the impact of AI in government around the world, with a focus on India.

Challenges and Roadblocks

Worldwide, there are a number of challenges that AI poses but the disruption that it would bring to the labour market is undoubtedly the one on top of everyone's minds. Opinions range from dire predictions about mass unemployment to those who claim that new positions or roles will be created. But truth be told, the potential impact of automation would depend not only on its scale but also on its distribution across the workforce and its timing. A detailed discussion on this issue features in Chapter 3.

Other challenges relate to issues of data privacy, cybersecurity, countering algorithmic bias, and preventing potential misuse of AI by criminals and terrorists. Many of these concerns cut across all uses of AI and it could have grave impact on governance, dissent, and freedom of expression. These issues are discussed in detail in Chapter 5.

Apart from the general challenges of AI, there are a few that are somewhat peculiar to India. The issue of capacity, for instance, is one of them. The private sector and more so the government need to build capacity, understanding, and competence for an effective implementation of AI-driven solutions.

Infrastructure is again a key concern in India. In the government sector, the infrastructural pre-requisite for the successful and cohesive implementation of AI-driven solutions have not been developed yet. The lack of internet penetration in rural areas and access to IoT devices are hindrances to adoption of AI-driven solutions in education of public delivery of services.

Compared to most large countries, funding is, perhaps unsurprisingly, a major challenge in India. Although the government has allocated some funding for Digital India and other programmes, its magnitude is quite low compared to that of the other big players such as the USA, China, or South Korea. It is also unclear how the allocated funds would be distributed across subsectors.

While regulation of AI is a budding area globally, India suffers from a virtual regulatory vacuum in this area. There is also legal and regulatory uncertainty on AI as only data privacy and data localization have attracted attention in terms of regulations. Sectoral regulators have also not indicated anything concrete. This may act as a disincentive to domestic and foreign investors or potential indigenous start-ups in terms of engaging with the sector.

Regulatory Frameworks

The regulation of AI is at a nascent stage in all countries across the globe. There are piecemeal legislations, policies, and standards that are currently governing the AI ecosystem. The European Union has led the way in regulations, the latest being the General Data Protection Regulation (GDPR). China and Russia have enacted strong laws in favour of data localization. There are also regulations regarding autonomous vehicles such as drones and self-driving cars.

India has just embarked on the path of framing regulations on data privacy. However, it has not yet considered other areas of AI for regulation. Some of the questions that India needs to grapple with before framing a comprehensive AI regulation are the following: who is liable if AI causes any sort of damage? Who owns the intellectual property that is created by AI? How should AI make choices and are there any categories of decisions which AI should not take?

A detailed discussion on the current regulatory landscape is presented in Chapter 5.

Roadmap for the Future

For India to take advantage of the AI revolution, it needs to develop an enabling ecosystem with digital infrastructure, improved access to and awareness about ICT, and has to focus on reskilling its labour force. It should also develop a strategy to re-negotiate its trade commitments in the WTO in such a manner that it can grow its nascent AI industry without being branded as excessively protectionist. There needs to be an adequate redressal mechanism for any misuse of AI or algorithmic biases that is accessible to all stakeholders. India also needs to widen the research base of AI to include members of other disciplines—lawyers, political philosophers, sociologists, anthropologists as well as software developers and corporates. This will ensure compliance with constitutional standards, fairness, and due process and also ensure that the benefits are distributed evenly among all strata of society.

The rest of this book will act as a curtain-raiser to the exciting world of AI to you, the interested reader. It will certainly not make you an expert of the field, nor will the figures, examples, and shape of things described here remain current for a particularly long period of time. Nevertheless, our hope is that it will provide some clarity and a broad overview of what may well be one of the greatest watersheds in the history of technological advancement of mankind. And that may very well be worth the effort.

2

What, Really, Is AI?

AI: Definition, Elements, Boundaries

Back in 1950, Alan Turing, one of the inventors of the modern-day computer, proposed a famous test named after him for machine intelligence. The test gauged if, on the basis of responses to queries, a human observer could distinguish between a human being and a machine. The term 'Artificial Intelligence' is generally believed to have been first used six years later at the amazingly over-optimistic Dartmouth Conference for (you guessed it) Artificial Intelligence, which hoped to crack the problem of thinking machines over a summer. More than six decades later, however, the essence of AI has remained the same—to get machines to do what is deemed to be intelligent in humans.

Intelligence, a notoriously slippery concept, is essentially a human quality: an ability to connect pieces of information to somehow recognize a pattern and to apply it on new information. In more technical terms, it consists of three steps: SENSE, perceiving or reading external signals; COMPREHEND, connecting them selectively to hypothesize a relationship between the various signals or pieces of data; and ACT, that is applying this model to respond to a set of new environmental signals. Even a young child does this instinctively—listening to words in a language; correctly associating them with objects or activities to infer meaning; and responding verbally or acting physically on the basis of this meaning. For a machine, however, doing this is anything but child's play. At its core, this is the challenge of AI.

There is, however, a striking similarity between an AI system and a child. Both need to LEARN to develop their intelligence. Comprehension abilities, whether in a child or in a machine, are essentially learnt, and consequently the *ability to learn* may well be taken as the defining characteristic of an intelligent system. Any program or software that predictively applies the same algorithm to new data—even if it processes terabytes of data as some weather predicting or rocket steering software do—would, by most opinions, fail to qualify as AI. If it is acting on a fixed rule, however nuanced and complicated, it is still just a program. To be intelligent, a machine, or often a system, has to be able to learn.

Consequently, the essential element of AI is machine learning. If a system is learning-enabled, it figures out a way to configure its own comprehension abilities and keeps on improving it, the same way as an intelligent human does.

It is, however, easy to over-emphasize the comprehension/learning part of the triad to the exclusion of the sensing and the acting part. As we shall soon see, machine learning itself is dependent on data collection/generation on a massive scale, and as a result the ability of the system rests entirely on (*a*) its ability to SENSE the environment, for instance, through image processing or speech recognition; and (*b*) the availability of massive quantities of data ('big data') to educate the system. Equally importantly, an AI system is ultimately only as impressive as the activity it produces—holding an intelligent and amazingly informative discussion like Siri, Alexa, or Cortana can; driving a vehicle through traffic like Google's driverless cars manage to do; or running 60 per cent of Wall Street's stock trades. In other words, the ACT side is just as important and fascinating technologically.

How does a machine learn? Turns out pretty much like children (or adults) do. Typically, a machine identifies an object such as a dog by being exposed to millions of pictures of it in various colours, angles, and postures. The in-built comprehension capability of the machine allows it to connect the images with one another so that in the next instance it can identify a dog's photo by finding it more akin to other pictures of dogs than anything else. This is called 'supervised learning' in AI lingo. The other kind, unsupervised learning, happens when the AI system perceives data without being told what it is seeing, so that it 'figures out by itself' the relationship between the various frames that it is observing.

This similarity between intelligent machines and humans is perhaps less surprising if we keep in mind that technically the learning process in AI systems is inspired by the functioning human brain—the network of neurons that gets activated in particular sequences and blocks. In machines this phenomenon is simulated by software to create artificial neural networks (ANNs) that function as the 'brain' of the system. In the last few years this ANN field has witnessed the application of a game changer—multi-layered ANNs—leading to what is known as 'deep learning', which is in degree and magnitude faster learning than what is enabled by traditional ANN technology.

The results of such learning can be startling. Google's AI application AlphaGo, which plays the notoriously complex boardgame Go, was taught the rules of the game and made to watch thousands of human Go games to figure out what works best. It soon ended up trumping legendary Go player Lee Sedol. Its next avatar, AlphaGoZero did not even need to watch human games, but *played* millions of games instead, against *itself*. AlphaGoZero mastered chess in less than four hours, particularly impressive given that it was not even specifically designed to be a chess bot. Some attribute its speedy success precisely to this fact: being a non-professional chess player it may have learnt something that had escaped people (and machines) who devoted their lifetimes to the game.

An interesting thing to note here is that what AlphaGoZero and AlphaGo learnt precisely about their games, however, is impossible to depict in a neat rule or formula. Much like humans, AI systems also learn things without knowing how they learn it. A master cyclist will be at a loss to explain exactly what routine of muscle movements he or she uses to ride the vehicle.

Almost all AI systems and applications that we currently have are what in AI parlance refers to 'narrow AI', intelligence used for a specific purpose. Say, for instance, a medical expert is looking at an X-ray to see if a patient has tuberculosis. The diagnostician can distinguish the X-ray of a tuberculosis patient from that of a healthy person from 'experience', that is, from the memory of many X-rays of both kinds that he or she has seen before. It is a relatively simple job for

AI to do, and it can usually outperform the human on accuracy and speed. In 2018, Babylon Health's AI product cleared the equivalent of the United Kingdom's general practitioner (GP) licence exam with a score of 81 per cent, well above the five-year average score of 72 per cent for human clinicians. Nevertheless, the application is a one-trick system excelling in a pointed task. In contrast, 'General AI', yet to be implemented, would be able to grapple with broad questions like ethical and philosophical debates and understand abstract concepts. Finally, *artificial super intelligence* is conceived to function at levels of complexity and abstraction well beyond the reach of the human mind.

Intelligence is also a layered concept. The Turing Test mentioned at the beginning of this chapter does not require a machine to be a conscious entity, it just needs to 'appear' to be intelligent. Google Assistant, which has received over half a million marriage proposals from Indians alone within months of its arrival, is at the end of the day just a search engine with speech and voice recognition abilities, and thus, technically speaking, a 'weak AI'. By contrast, 'strong AI' resembles human thought by a few more notches, being able to make logical connections between data points at a level or two deeper than at the surface. AlphaGoZero would certainly qualify as 'strong AI'.

While these distinctions are good to bear in mind, it is also important to appreciate that the boundaries between them are essentially fuzzy, and that it is best to think of them as regions in a continuous spectrum rather than as clearly segregated pigeonholes. From a simple deterministic program to 'smart technology' that can read its environment and select between a choice of pre-programmed responses, to machine learning–enabled systems that can evolve with data, intelligence has been deepening and progressing, and pushing the envelope at multiple levels. So much so that futurists today predict a world three short decades ahead where humans would be able to connect to machine intelligence wirelessly breaking the ultimate barrier between humans and machines to create AI-enhanced human intelligence.

A Brief History of AI

The 1950s can have an undisputed claim over the early development of AI. In 1952, four years before the notoriously over-ambitious summer conference at Dartmouth, Arthur Samuel developed the first computer program that could learn on its own. Three years later Herbert Simon and Allen Newell created Logic Theorist, the first artificial intelligence program, capable of solving complex mathematical problems.

The year 1958 witnessed John McCarthy creating the programming language LISP—the workhorse of AI research.

By 1961, Unimate, the first industrial robot, started working in the New Jersey plant of General Motors, and in 1965 Joseph Weizenbaum built ELIZA, an interactive program that could carry on a conversation in English on almost any topic. A year later, Shakey, the first mobile general purpose robot capable of breaking down complex orders into smaller tasks and performing them made its appearance. It was a product of research funded by the military of the United States of America (USA) at the Artificial Intelligence Center of Stanford Research Institute (now SRI International). Programmed in LISP, Shakey was developed between 1966 and 1972 and could carry out simple tasks such as moving from room to room, operating switches and pushing objects without any remote human control. Shakey received wide publicity and caught the imagination of much of the world.

Notwithstanding Shakey's popularity, interest in and funding of AI experienced major swings with long droughts, including one ushered in by the 1966 ALPAC committee report in the USA.[1]

[1] The US government formed ALPAC (Automatic Language Processing Advisory Committee) in 1964 to

The US government did invest up to USD 20 million in AI research as it sought to use machine intelligence to auto-translate Russian documents during the Cold War. These efforts hit a wall in the 1960s after achieving only meagre progress and the funding dried up. The setback to funding continued in the 1970s when a speech recognition project at Carnegie Mellon University failed prompting the Defense Advanced Research Projects Agency (DARPA) to cancel a multimillion dollar research grant. Similar debacles occurred in the 1980s and 1990s as LISP, an early family of computer programming languages, failed to take off and expert systems gained little commercial traction.

In the UK too AI research was mostly abandoned in the 1970s after the Lighthill 'report' (a survey paper in reality), which was commissioned by the British Science Research Council to evaluate AI progress, cited the utter failure of AI research to achieve its big objectives.

---

evaluate the progress in computational linguistics in general and machine translation in particular. It was headed by John R. Pierce and included seven scientists. Its report, issued in 1966, was very sceptical of research done in machine translation so far, and emphasized the need for basic research in computational linguistics. This was one of the reasons the US government reduced its funding of the subject.

Nevertheless, machine learning continued to progress almost in step with faster computing over the years. In 1978 the XCON program developed at Carnegie Mellon University helped in configuring computers based on buyer demand; a year later the Stanford Cart managed to cross a chair-filled room without human support in about five hours. In 1986 a team led by Ernst Dickmanns built the first driverless car at Bundeswehr University at Munich that could drive up to a speed of 55 mph on empty streets.

Impressive first steps all these doubtless were, but machines matching human intelligence was still a matter of science fiction. That changed in 1997 when IBM's Deep Blue, a chess-playing computer, succeeded in beating the reigning world chess champion Gary Kasparov. AI had finally come of age.

The two decades since this landmark event have witnessed an explosion in technological development in AI at a pace much faster than before. Also, AI is being increasingly put to use in solving real world problems. For example, the new millennium brought forth the development of two robots, Kismet developed by MIT's Cynthia Breazeal that could recognize and simulate emotions and Honda's iconic humanoid Asimo robot that could walk at a human pace and deliver trays to customers in a restaurant.

In 2009 computer scientists at Northwestern University's Intelligent Information Laboratory developed Stats Monkey, a program that could write sport news stories without human intervention. In the same year, Google started its secret driverless car project. In five years, it produced one that could pass the self-driving test of a US state. In 2011 a neural network system managed to win the German Traffic Sign Recognition competition with 99.46 per cent accuracy beating humans at 99.22 per cent. In the same year IBM's Watson, a natural language question answering computer, defeated two former champions in the popular US TV show *Jeopardy!* As mentioned earlier, five years later in 2016, Google DeepMind's AlphaGo succeeded in defeating Lee Sedol, a world champion of the incredibly complex game of Go. Months later, another application was developed that could beat this virtual champion 100–0!

Public funding for AI has revived and in the last decade, much of the funding has come from DARPA's Cyber Grand Challenge, a competition with prize money, and the European Union's EU–FP7 technology funding programme.

The growth in both interest and application of AI since 2000 has been truly phenomenal. According to the AI Index report 2017, the number of active AI start-ups in the USA has risen 14-fold. Scopus reports a 9-fold increase in the annual number of published scientific papers on AI since 1996. Student enrolment in AI courses in major universities have risen five to ten times over the period.

What lay behind the increase in the pace of development of AI that we have witnessed in the new millennium? A convergence of three factors: quantum change in processing power; drastic reduction of data storage costs; and a veritable explosion of data. Processing power has been rising at a rapid pace over several decades (Moore's law) but a significant change came with the transition from central processing unit (CPU) technology to graphical processing unit (GPU) technology. Now the move to Google's newly developed tensor processing unit (TPU) technology holds the promise of bringing in more drastic change. In parallel the costs of storing data has been falling at an incredible pace, from about half a billion US dollars to store a gigabyte (GB) in 1980 to about 2 cents in 2017. This has facilitated an ever-growing tsunami of data around the world. In 2016 the total amount of data generated exceeded 16 zettabytes (1 zettabyte = 1 trillion gigabytes).By some projections, this is expected to grow by ten times over the next decade.

This third element, data explosion, facilitated significantly by the other two, is the direct driver of the rise of AI. Ultimately AI and machine learning feed on data, GB of it. An AI system is about as smart and as accurate as the amount of data it has had to munch on. As the world moves to a state where every second generates data from digital cameras and digital conversations that is now affordable to store, AI technology gets its supply of fodder to train on. Availability of open source software and parallel computing using computer networks on the cloud have contributed massively to the onward gallop of AI as well.

While the USA has doubtless been the cradle of early AI development and continues to be the main source of key AI advances, in the last decade China has demonstrated significant capability, application, and growth in the area, so much so that today China surpasses the USA in annual scientific papers published on the subject, though the USA still dominates massively when it comes to citations—the time honoured measure of impact of research. Given its size, China appears to have an advantage in its generation of and access to data to train AI systems on. During 2010–15 China filed around 8,000 AI-related patents, a growth rate of 190 per cent, faster than any other country. While the growth of the sector is propelled by the government's explicit objective of dominating AI, the real movers in the sector have been a slew of companies including Baidu, Alibaba, and, perhaps most notably, Tencent.

It is interesting to note that the beginnings of AI in India can actually be traced at least to the early 1960s, barely a few years after its global beginnings. It is instructive to sketch the journey of development of computing and AI in India since Independence. A few distinct 'stages' separated by prominent break points become apparent quickly.

Stages of Development of Computing and AI in India[2]

1955–70: India established the Indian Institutes of Technology (IITs) and started the design and production of computers. In fact, the first course on AI was introduced at IIT Kanpur in the late 1960s when Professor H. N. Mahabala returned from the Massachusetts Institute of Technology after spending a year there and having interacted with Professor Marvin Minsky. In 1963, the government-appointed Bhabha Committee realized the importance of electronics and

[2] V. Rajaraman, 'History of Computing in India: 1955–2010', Indian Institute of Science, 2012, available at https://history.computer.org/pubs/2012-12-rajaraman-india-computing-history.pdf, last accessed on 18 February 2020.

computers in national development and suggested the establishment of the Department of Electronics (DoE) in the government. The DoE was established in 1970, which was the first break point.

1971–8: The DoE stressed self-reliant indigenous development of computers, and a company called the Electronics Corporation of India Ltd (ECIL) was financed to design, develop, and market computers using components primarily made in India. ECIL made computers called TDC 312 and TDC 316, which were similar to the PDP series computers made by the Digital Equipment Corporation of the USA. The DoE also initiated many R&D projects with assistance from the United Nations Development Programme (UNDP). The second break point came in 1978 when IBM closed its operation in India due to differences with the government. The new Janata Party government decided to open up computer manufacturing to the private sector and a number of companies started making minicomputers using imported microprocessors. UNIX was the Operating System of choice.

1979–86: The sector received a boost in 1984 and 1986 when the Rajiv Gandhi-led Congress government removed numerous controls on the industry and on imports. This led to a sharp reduction of price and a speedier spread of computer use. In 1986 software companies were allowed to import computers at reduced import duty rates. Software development was recognized as an industry deserving many tax concessions. The success of the computerized ticket reservation system of the Indian railways made the public aware of the relevance of computers. Around the same time, India saw the telecom revolution under the leadership of Sam Pitroda and the Centre for Development of Telematics (C-DOT).

Research on AI also started evolving with projects such as Machine Translation for the Indian languages project at IIT Kanpur and the Optical Character Recognition project at Indian Statistical Institute (ISI), Kolkata. In 1986 the DoE, with assistance from UNDP, started the Knowledge-Based Computing Systems (KBCS) project as part of the Indian Fifth Generation Computer Systems research programme focusing on

- intelligent man–machine interface;
- knowledge-based processing and management function;
- problem solving and inference making function;
- development of parallel processing platforms for KBCS.

The goal of KBCS was to develop a state of the art computer programming environment. The nodal centres under KBCS included the Centre for the Development of Advanced Computing (CDAC), DoE, Indian Institute of Science (IISc), IIT Madras, ISI, Tata Institute of Fundamental Research (TIFR) , and National Centre for Software Technology (NCST). The KBCS applications included expert systems for government administration, engineering and medical applications, intelligent tutoring-authoring systems, computer vision system applications, and KBCS applications in and for ancient Indian sciences.

1987–97: The third break point came in 1991 when India was forced to open its economy and reduce controls on local manufacturing companies. Software and software services companies formed the National Association of Software and Services Companies (NASSCOM) which successfully lobbied with policymakers and obtained many tax and other concessions. One of the major initiatives taken by the DoE at this time was the establishment of software technology parks (STPs) with satellite communication links, which enabled Indian software companies to develop software applications on their international clients' computers from India.

1998–2012: The fourth break point was in 1998 when the new government under Atal Bihari Vajpayee declared 'IT as India's tomorrow', and took a number of proactive measures to promote software companies.

An IT task force was appointed to recommend changes in the policies of the government. Measures were taken to give a tax holiday on the export earnings of the Indian software services companies for 10 years and import duty was exempted on computers and software packages imported for exporting software. Multinational companies were welcomed to set up software development and R&D centres.

2013–19: AI started moving from the laboratories to commercialization as companies started using AI in designing a variety of products and services, increasing productivity, efficiency, and reliability. AI powered start-ups were formed attracting venture capital funding. Also, industry funded research projects witnessed a boost.

At the Boundary of AI—Big Data, Blockchain, Cryptocurrencies, Internet of Things

Recent years have witnessed an explosion of several new potentially revolutionary ideas with vast usability that often get labelled as part or extension of AI. Blockchains are everywhere today, finding new applications every day. Millions are being won or lost with the gyrations in the value of cryptocurrencies. The internet of things (IoT) holds the promise of a truly sci-fi world. And big data, used by bankers to marketers alike, already seems passé. It helps to define and delineate these concepts though and make clear their relationship with AI.

Big data may very well be the best place to start. Data lies at the heart of all computing and massive amounts of data have been managed by computer systems for several decades now. The term 'big data' applies to situations where the data in question is so large, and grows so fast that standard ways of data management, that is, storage and analysis, are simply not suitable to handle the deluge. For instance, in stock trading, the New York Stock Exchange generates about one terabyte (TB) (about 1,000 GB) of data per day. If that seems impressive, a single jet engine can generate about ten times as much data in 30 minutes, and even that pales before the 500+ TB of data that Facebook adds every day.

Handling big data is definitely not an easy task but if one can manage it, the benefits can be immense. Analysing and connecting big data can give unparalleled insights into the mind of the consumer and her likely future behaviour. Businesses increasingly use it to push products and advertisers around the world vie for this ability.

Big data is the cradle of AI in the sense that training a learning application essentially requires its use. In and of itself, however, big data cannot be considered part of AI, but its role is more of being the fodder for a learning application to hone its 'intelligence'. The relationship is more complex than this, however. An important way in which big data analysis claims to differ from standard statistics is in its treatment of finding patterns from data itself without imposing any models or hypotheses on it. To the extent this works—and this may differ vastly across situations—the analysis of big data may also be seen as AI itself as it boils down to identifying patterns in data.

If the use of big data has already become fairly standard in most large business organizations, another frequently talked about but less frequently understood innovation that is finding rapid adoption and use across sectors is blockchain. If big data is about quantum of information, blockchain is about securing information by sharing copies of it within a network of computers or computer systems. The essential idea is to maintain an identical ledger of transactions with every participant in an extended (and fast growing) network; this makes the records interconnected, which aids in verification, making hacking next to impossible as a necessity is created to hack every system in the network in order to produce an identical result at all nodes.

Blockchain technology caught the attention of non-technical people mostly through the emergence of a new genre of 'currency' called cryptocurrency (which we will discuss later) but has since found rapid and increasing usage in almost all sectors where information security is key, primarily in the financial sector. In India, ICICI Bank has been among the early movers in adopting blockchain technology in the area of banking.

Blockchains are obviously critical in ensuring authenticity in environments with transactions. Examples of this can range from settings such as land ownership records—the single largest source of court cases in India—to electronic payment transactions to checking the authenticity of high-value artistic work. Strictly speaking, they are distinct from AI in the sense that they are mostly about sharing information and maintaining records securely rather than 'learning'. However, interesting examples of convergence do occur. Doc.ai, for instance, provides AI solutions to patients rather than have human doctors treat them. Its bank of medical information comes from a network of subscribers whose medical data is protected using blockchain technology. Therefore, blockchain technology can provide an important support function for AI where the source of data needed for the AI application to 'learn' comes from a network of individuals or computer systems.

Blockchains became known to most people because of the far more exciting technology they support:

cryptocurrencies. Cryptocurrencies were made famous by Bitcoin launched in 2009. Since then, over 1,600 cryptocurrencies including Ethereum and Ripple have appeared with no central bank issuer. They are currencies accepted by a network of users using blockchain technology. The excitement about them of course stems from their sharp rise and occasional fall in valuation in terms of real money. In the decade since the emergence of Bitcoin, the combined market valuation of cryptocurrencies has crossed the USD 100 billion mark, higher than the GDP of 125 countries. Bitcoin value crossed the USD 6,000 mark in 2018. And yet economists remain divided about their viability and sustainability in the long run as cryptocurrencies are backed not by a central bank of a sovereign country but a network of users. Markets, however, remain excited and new cryptocurrencies are regularly entering the field.

The direct connection between cryptocurrencies and AI is hard to find. They are distinctly different innovations though the latter is claimed to have been used in several trading strategies for cryptos, which could have been used for trading any liquid asset.

If big data is the recent past and blockchains and cryptocurrencies the present of a fast-evolving new tech world, then the internet of things (IoT) can still be thought of as the future, though in a rudimentary way it has been in existence for decades now. At its core, IoT refers to appliances communicating with one another and humans using embedded electronic technology. To the extent this involves reading and interpreting signals, IoT is much closer to AI in principle, though the defining element here would be in the nature of the response of the smart device. In case the response is pre-programmed—like an RFID tag in a car communicating with a toll gate reader—it would technically not be AI, whereas if the interaction can possibly evolve over time through detection of patterns then it fulfils the condition of being 'intelligent'.

Evolution in big data, blockchain, and IoT together with giant strides in robotics, which of course is part of AI itself, have all been happening together with rapid advancement in machine learning in the last two decades. In a narrow, exact, technical sense, AI should be defined as 'learning' processes, but loosely speaking, all of these related but distinct developments often get clubbed with narrow AI in a broader definition of the term.

3

AI and Industry in India

Overview

Most believe that AI will transform almost every aspect of our lives. The current debates have moved to questions of how, when, and where the impact of AI will hit the hardest. AI is already having an impact across virtually every industry be it IT, manufacturing, retail, health care, financial services, education, or media. It ranges from helping employees at transportation companies predict arrival times to predicting toxins in grains of food to helping scientists learn how to treat cancer more effectively.

While businesses largely anticipate a positive impact on growth, productivity, innovation, and in some cases job creation, there are challenges such as biases in algorithms, lack of data storage space, need for massive skill upgradation, and the drawbacks of the one-track mind of specialized AI that need to be taken into consideration. Competitive strengths are likely to get redefined as well. Each country is at a different stage in the AI continuum. Where does India fit in in this? How prepared is it to tackle the opportunities and challenges that AI will inevitably pose?

Studies show that India has a significant workforce equipped with AI skills; its companies are among the early adopters of AI and it ranks third in research on AI. But it also lags far behind in private sector investment in AI. The high initial cost of implementing AI-based solutions deters start-ups from deploying AI technologies. Moreover, there are multiple legal and privacy-related implications of utilizing citizen data. There is also a severe lack of clean data repository in India since most of the records are still in paper-based forms. Added to all this are the concerns over the possibility of job losses due to automation and implementation of AI in organizations.

While the present scenario in India's AI ecosystem is a mixed bag, there are winds of change in the right direction. The government has recently taken steps to spur AI research and adoption and allocated a sizeable fund for the purpose. The biometric and demographic data captured under India's Aadhaar programme may be used as a base by the government to train any in-house algorithms. Over 30 per cent of companies were planning to expand their AI budgets by more than 10 per cent in 2019. There is a strong probability that jobs displaced by AI will be substituted by new jobs that require a radically different skill set. AI will allow humans to get more involved in tasks that need higher specialization and critical thinking.

This chapter attempts four things: first, to outline India's position in the world in terms of its AI preparedness focusing on investments, human resources, and adoption of new AI-based technologies; second, to identify a few key sectors/industries that are changing most rapidly due to disruption by AI and examine the developments taking place there; third, to identify the challenges as well as the roadblocks that are likely to be thrown up due to the adoption of AI in several sectors; finally, to present a sketch of the opportunities that exist in sectors not yet showing an uptick in the adoption of AI.

India and the World: Investment, Human Resource, Adoption of AI

Investments

Where Is India in Terms of Investment in AI?

The government has been a key funder of AI over the past half a century but the decadal ebb and flow of investment in AI has primarily been guided by the possibility of practical AI applications. The US government funded AI research in a big way between 1950 and 1970, after which it slumped when the government did not see the desired result in the application of AI. The field remained fallow for more than a decade till the 1980s when university researchers developed 'expert systems'—software programs that assess a set of facts using a database of expert knowledge and then offer solutions to problems. Around this time, the first computer-controlled autonomous vehicles also began to appear. However, the revival of AI was short-lived. Interest in AI boomed again in the twenty-first century as advances in fields such as deep learning, underpinned by faster computers and more data, convinced investors and researchers that it was practical—and profitable—to put AI to work.

Is this boom here to stay? According to a report by McKinsey, the answer is in the affirmative as AI is finally starting to deliver real-life business benefits. The ingredients for a breakthrough are in place—computer power is growing significantly, algorithms are becoming more sophisticated, and, perhaps most importantly, the world is generating vast quantities of the fuel that powers AI data: billions of GB of it every day.

If government funding has played a key role in developing AI around the world, private investment has not lagged far behind. It has, in fact, assumed pivotal status in recent years. Investment in AI by large corporations as well as funding from venture capital and private equity funds is growing rapidly. It is dominated by tech giants and digital native companies such as Alphabet (Google's parent company), Baidu, Apple, Facebook, Microsoft, IBM, and Amazon, which develop the inputs needed to enable AI applications— powerful computer hardware, increasingly sophisticated algorithmic models, and a vast inventory of data. These tech giants spent USD 20 to 30 billion on AI in 2016, 90 per cent on R&D and deployment and 10 per cent on AI acquisitions. Venture capital and private equity financing, grants, and seed investments in AI start-ups also grew rapidly, albeit from a small base, to a combined total of less than USD 2 billion in 2013 to over USD 6 billion in 2017. Machine learning, as an enabling technology, received the largest share of both internal and external investment.

The USA is clearly the pioneer here. It boasts of the strongest ecosystem for AI in terms of funding, number of companies, and global reach. About 40 per cent of all AI companies are based in the USA. Its leadership is a result of a mature, well-financed, and thriving ecosystem in the Silicon Valley and New York/Boston metropolitan area. Over 16 governmental agencies support AI companies financially and politically (including DARPA, the Central Intelligence Agency [CIA], and the National Security Agency [NSA]), and it boasts of leading universities (such as Stanford and MIT) as well as very strong corporate research facilities (such as Google DeepMind).

USA is followed, at a distance, by China, Israel, and the UK. Approximately 13 per cent of all AI companies are based in China, 12 per cent in Israel, and 8 per cent in the UK.

China's commitment to AI is total. AI is the focus of China's development agenda and it aims to be a leader in the field in the next five years. Its ambitious Next Generation AI Development Plan, published in 2017, sets a number of clear targets: to reach the same level in AI as the USA by 2020, to become the world's premier AI innovation centre by 2030, and to build a domestic AI industry worth USD 22.2 billion by 2020 and USD 59.1 billion by 2025. The government has budgeted investments of USD 5 billion and private entities another USD 2 billion in AI efforts in start-ups, universities, and in attracting American firms.

This emphasis is beginning to show in the statistics regarding the growth of AI. Although the USA remains the top country for AI investment with a share of more than 25 per cent of global R&D funding, China is close at 22 per cent.. While data on private sector investments in China is often unreliable, according to one estimate, China accounted for a greater share of global private sector funding than the USA in 2017. In terms of equity funding of AI start-ups in 2017, China accounted for 48 per cent of world funding, USA 38 per cent, and the rest of the world chipped in with the remaining 13 per cent.

The market for start-ups in China is very well-financed, and valuations are even higher than in the Silicon Valley. It has the strongest growth based on published academic papers, a surprisingly high number of AI start-ups, and research centres have been set up in Beijing and Tianjin. However, the USA has already invested heavily in AI since 1999 while China is now catching up. In 2017, out of the total venture capital invested in AI (USD 28.9 billion), the share of the USA was at USD 14.8 billion, while China stood at USD 9.6 billion.

In contrast, India had 82 AI start-ups in 2017 (about 3 per cent of all AI start-ups globally) as compared to Israel's 342 and China's 383, not to speak of the USA's 1,353. This placed India at the ninth position globally, but it was the only non-OECD country, barring China, to feature in the world top 10. According to NASSCOM, the AI start-up pool is expanding rapidly at over 50 per cent compounded annual growth rate since 2013. The key segments are enterprise, marketplace, health-tech, ed-tech and fintech. But the range of application is quite stunning. Among emerging start-ups, AnsweriQ provides an AI-enabled customer support ticket management solution, while AskSid and Wysa are AI-based chat bots for business solutions and emotional support respectively. More mature start-ups range from SigTuple, applying AI-powered analytics to visualize medical data, and Flutura, combining AI and IoT to bring predictive analytics to manufacturing, to Zendrive, an AI-based platform to monitor risky driving behaviour and to trigger real time alerts and Active.Ai, a conversational banking platform for financial institutions.

Notwithstanding this impressive array, India lags quite a bit behind in terms of private sector investment in AI compared to the leading AI ecosystems globally, especially in the USA and China. However, 2018 has witnessed a jump in investment and deal activity around intelligent automation and artificial intelligence, machine learning, and big data. Start-ups scaled an all-time high in raising capital in 2018, up more than three and a half times from the 2017 figure. Start-ups with operations in India raised over half a billion US dollars in funding rounds. This includes start-ups with investment at varying stages of development, from pre-seed to well-funded companies. Automation Anywhere based out of California and India bagged the biggest cheque of USD 300 million from SoftBank

Vision Fund. It is believed that the growth story is set to continue as well. The size of the AI market in India is projected to grow 30 times to USD 90 billion in 2025 from just over USD 3 billion in 2016.

India's AI start-ups are, unsurprisingly, younger. According to the sample of over 2,300 global AI start-ups of Bengaluru-based incubator Excubator, about 10 per cent of AI-based firms have witnessed exits—208 acquisitions and 28 IPOs. In contrast, India has had only two exits of significance in 2017: Apus acquired Siftr, which used computer vision to curate images from user-generated content, and Google acquired Halli Labs, which was developing AI-powered speech and vision products. However, more exits have taken place in 2018 and 2019. For example, Capabiliti was acquired by Peoplestrong, Cube26 by PayTM, Ableplus by OYO, Kogentix by Accenture, Liv.Ai by Flipkart, Tapzo by Amazon, Sigmoid Analytics by Google, Int. Ai by Walmart Labs, Shieldsquare by Radware, and Kint.Io by Swiggy.

How does this compare with China? India seems to be scrambling to make up for lost time, which has led to some significant developments in the sector. Policymakers appear to have woken up as well. The government's think tank, NITI Aayog, has developed a strategy document positioning India as the AI 'garage for emerging and developing economies'. The government has budgeted about INR 3,073 crores in 2018 and there have been moves to focus on R&D and investment in AI.

Private sector giants are not missing out on the action either. Indian telcos Bharti Airtel and Reliance Jio have set up labs to carry out AI research. Infosys, Wipro, and other such IT giants have begun delving into the investment side of the AI market, making equity investments in many AI-based start-ups. Multinational Corporations (MNCs) such as NVIDIA, Microsoft, and Google have also set up R&D labs in India to benefit from the available skilled human resource.

Challenges abound as well. There is a severe lack of a clean data repository in India since most records are still restricted to paper-based forms. Aadhaar, which over 99 per cent of India's population has signed up for, provides a silver lining. With over 1.19 billion records in its database, the increased adoption of Aadhaar might provide a base for the government to train any 'in-house' algorithms. Industry experts point that the lack of network effect, lack of focus on deep tech, and relatively low investment in AI start-ups by risk capital providers as compared to the USA and China are among other issues.

By now it is amply clear that India has a key advantage in terms of availability of AI-specialized human resources but it is quite a distance behind China, which is clearly far ahead in every other way and is focused on wresting the crown of the Fourth Industrial Revolution from the USA.

Human Resources

How Ready Is the Indian AI Workforce?

India certainly has a size advantage. According to a report by LinkedIn,[1] India ranks among the top three countries in AI skills penetration after the USA and China, ahead of Israel and Germany. These skills include but are not limited to neural networks, deep learning, natural language processing, computer vision, robotic process automation, speech recognition, machine learning, as well as competency in actual tools such as Weka and Scikit-Learn.

The report cites primarily three drivers: a rise in data and programming skills that are complementary to AI; skills to use products or services that are powered by data such as search engine optimisation for marketers; and interpersonal skills.

Notwithstanding the relative position, the need for reskilling in India is equally daunting. A large segment of India's half-a-billion-strong labour force, mostly

[1] Igor Perisic, 'How Artificial Intelligence Is Already Impacting Today's Jobs', LinkedIn Economic Graph, 17 September 2018.

engaged in agriculture, manufacturing, and low-skilled white-collar jobs in the services sector, rank seriously low on the AI skills ladder. There is a severe need for reskilling to avoid loss of livelihood with the unfolding of the Fourth Industrial Revolution. According to a NASSCOM report, about 40 per cent of India's total workforce has to be reskilled over the next five years to cope with emerging trends such as AI, IoT, machine learning, and blockchain.

The Modi government in its second-term is actively considering adding reskilling in AI, IoT, and machine learning as part of its Skill India initiative. It has identified over half a dozen sectors in which a dedicated curriculum for reskilling will be developed based on demand from these sectors.

According to a World Economic Forum report on the future of jobs, the window of opportunity for proactive management of this transformation of the workforce is closing fast, and businesses, governments, and workers must proactively plan and implement a 'new vision for the global labour market'.[2] The report identifies four key drivers of change that will impact business growth: high–

[2] World Economic Forum, 'The Future of Jobs 2018', Insight Report, 2018, available at http://www3.weforum.org/docs/WEF_Future_of_Jobs_2018.pdf, last accessed on 18 February 2020.

speed mobile internet, artificial intelligence, big data analytics, and cloud technology. Many companies also expect that they will have to significantly modify how they produce and distribute by changing the composition of their value chain or modifying their geographical base of operations. This would be decided by a range of factors including availability of skilled local talent, labour costs, flexibility of local labour laws, industry agglomeration effects, or proximity of raw materials.

Furthermore, the report finds that in India 54 per cent of workers across 12 industries would need to be reskilled by 2022. While 35 per cent of the workers need at least six months of reskilling, one in ten would need over a year of training to be ready for the workplace of the future.

How does India stack up vis-à-vis other countries in terms of the need to reskill its workforce? Taken as a percentage, the reskilling needs of the countries selected in the report do not seem to vary hugely across countries. But given India's population, in absolute numbers it would still be a massive challenge.

What Could Be the Impact of AI on Jobs in the World and in India?

While the transformative impact of AI on every industry is indisputable, the most unsettling question regarding AI is about the impact it would have on the labour market. Dire predictions about mass unemployment are countered by those who opine that new positions or roles will be created. The potential impact of AI would depend not only on its scale but also on its distribution across the workforce and its timing and speed.

Several studies aim to estimate the proportion of current jobs that could technically be automated in the future. These estimates aim to provide a sense of the scale of potential transformation that could be enabled by technology, specifically the proportion of current employment for which the immediate first order effect of AI adoption could be radical transformation or displacement. Some of these studies rely on assessments of the technical automatability of existing work tasks, coupled with data analysis to investigate which job characteristics (for example, a requirement to interact with customers) are correlated with the assessed automatability. Others are based on expert opinions.

A two-year study from McKinsey Global Institute suggests that by 2030, intelligent agents and robots could eliminate as much as 30 per cent of the world's human labour. Depending upon various adoption scenarios, the report estimates that automation will displace between 400 and 800 million jobs by 2030, requiring as many as 375 million people to switch jobs categories.[3]

A Pew Research Center study asked 1,896 experts about the impact of emerging technologies and found that about 48 per cent of these experts envision a future in which 'robots and digital agents have displaced significant number of both blue and white collar workers with many expressing concern that this will lead to vast increases in income inequality, masses of people who are effectively unemployable, and breakdowns in the social order'. The other half of the experts surveyed expected that technology would not displace more jobs than it creates by 2025. It is important to note that this group also anticipated that many jobs currently performed by humans would be automated by 2025 but they chose to repose faith in human ingenuity to create new jobs, similar to the events post the Industrial Revolution.

---

[3] McKinsey Global Institute, 'Jobs Lost, Jobs Gained: Workforce Transitions in a Time of Automation', December 2017, available at https://www.mckinsey.com/~/media/ mckinsey/featured%20insights/Future%20of%20Orga- nizations/What%20the%20future%20of%20work%20 will%20mean%20for%20jobs%20skills%20and%20wages/ MGI-Jobs-Lost-Jobs-Gained-Report-December-6-2017. ashx, last accessed on 18 February 2020.

Another study by PwC paints an optimistic picture, predicting that AI would encourage a gradual evolution in the job market that will be positive with the right preparation. It suggests that a 'centaur', which is a human and AI working as a team with the human taking advice from AI but having the power to override it, will be the real key to success. For example, an AI chess grandmaster can be defeated by a centaur. It also predicts that functional specialists, not techies, will decide the AI talent race. Therefore, an AI application to support asset management decisions would need economists, analysts, and traders on a continuous basis.

More detailed analyses showed that the impact of automation on jobs would probably be in the range of 14 to 54 per cent. A 2013 study by Frey and Osborne, researchers at the University of Oxford, found that 47 per cent of US workers have a high probability of seeing their jobs automated over the next 20 years.[4] Using the same methodology as Frey and Osborne, it

[4] Carl Benedikt Frey and Michael A. Osborne, 'The Future of Employment: How Susceptible Are Jobs to Computerisation?', 17 September 2013, available at https://www.oxfordmartin.ox.ac.uk/downloads/academic/The_Future_of_Employment.pdf, last accessed on 18 February 2020.

was found that on average 54 per cent of EU jobs are at risk of computerization, especially those which are in low-wage, low-skill sectors traditionally immune from automation.[5] On the other hand, researchers at the Organisation for Economic Cooperation and Development (OECD) focused on 'tasks' rather than 'jobs' and found fewer job losses. Using task-related data from 21 OECD countries, they estimated that 14 per cent of jobs were highly automatable and another 32 per cent have a significant risk of automation. While they predicted lower job loss estimates, they warned that low-qualified workers were likely to bear the brunt of the adjustment costs as the automatability of their jobs was higher compared to higher-qualified workers.[6]

While the estimation of automation may vary, it is clear that AI technology will change the business world in three aspects: automation, intelligence, and

[5] Jeremy Bowles, 'Chart of the Week: 54% of EU Jobs at Risk of Computerisation', 24 July 2014, available at https://bruegel.org/2014/07/chart-of-the-week-54-of-eu-jobs-at-risk-of-computerisation/, last accessed on 18 February 2020.

[6] M. Arntz, T. Gregory, and U. Zierahn, 'The Risk of Automation for Jobs in OECD Countries: A Comparative Analysis', OECD Social, Employment and Migration Working Papers, no. 189, OECD Publishing, Paris, 2016.

creation. In most sectors, it will make some jobs, especially those requiring low skills, redundant while possibly increasing efficiency and creating new types of jobs. However, occupations that are likely to grow disproportionately are those needing high education, although some middle-education occupations are also likely to grow.

However, changes are likely to occur at both ends of the skill-intensity spectrum. A few high-skill sectors are at risk, for example, financial specialists (including fraud examiners, risk management specialists, financial quantitative analysts, and investment underwriters), and to a lesser extent, health technologists and technicians (such as radiologic technicians). In fact, a 2019 study by Brookings, which generated a measure of every occupation's varying levels of exposure to AI applications in the near future, shows that white collar jobs (better paid, better educated professionals) will be most affected by the new AI technologies. But it would also impact low-skill sectors such as manufacturing and production (assemblers, fabricators, food processing workers, and other hospitality occupations). Similarly, growth is likely to be seen in high-skill sectors such as creative, digital design, engineering, architecture, medical, education occupations as well as low-skill ones such as animal care and service workers, and personal appearance workers.

High job losses in some Western democracies could lead to grave socio-economic inequality, bringing in its wake potentially violent political outcomes and rise of authoritarian rulers and policies to stave off civil chaos, much like they did during the Great Depression.

What will the future of jobs look like in India? There are two key factors that pose a challenge: (*a*) approximately 17 million enter the workforce in India year on year while only 5.5 million jobs are available; and (*b*) the speed and scale of the disruptions impacting the way we work and live. According to a report by Ernst & Young (EY), the workforce in 2022 would look very different from today—about 9 per cent of it would be deployed in new jobs that do not exist today; 37 per cent would be deployed in jobs that require radically changed skill sets; and the remaining 54 per cent jobs would fall under unchanged job category.[7] The distribution would also vary substantially across sectors. In IT for instance, about 10–20 per cent are likely to work in new jobs and 60–65 per cent in those with changed skill sets, while for retail the figures are likely to be in the 5–10 per cent and 20–25 per cent

---

[7] NASSCOM, FICCI, and EY, 'Future of Jobs in India: A 2022 Perspective', 2017, available at http://ficci.in/spdocument/22951/FICCI-NASSCOM-EY-Report_Future-of-Jobs.pdf, last accessed on 18 February 2020.

range only. A vast range of new jobs are expected to come up by then. These may include roles such as 3D modelling engineer and designer, cloud architect, data scientist, automobile analytics engineer, e-textiles specialist, robot programmer, and blockchain architect.

What Careers Will Be Most in Demand?

The World Economic Forum's (WEF's) 2018 report estimates that machines and algorithms in the workplace will create 133 million new roles, but cause 75 million jobs to be displaced by 2022. This means that the growth of artificial intelligence could create 58 million net new jobs in the next few years. The fastest growing job opportunities across all industries include data analysts, software developers, and social media specialists along with jobs that require 'human skills' such as sales and marketing, innovation, and customer service. But jobs such as data entry, payroll, and certain accounting functions could disappear.

There are certain long-term implications of the rapid advance in AI technology. Those without appropriate skills are probably going to get left behind, which makes it essential to focus on developing strong AI skills as early as possible. Working with AI, machine learning, and deep learning involves knowledge of a variety of skills. While someone with programming skills is an obvious candidate, an average employer would also want someone with an understanding of a certain kind of maths, namely linear algebra and statistical methods. Along with maths, a good understanding of large data sets is essential.

According to a study by Accenture in 2017,[8] three new categories of AI-driven business and technology jobs will be created. The study categorized these under three labels: trainers, explainers, and sustainers. Humans in these roles will complement the tasks performed by cognitive technology and ensure fairness, transparency, and auditability.

- Trainers: This category of jobs will need human workers to teach AI systems how they should perform. For example, customer service chat bots need to be trained to detect the complexities and subtleties of human communication.
- Explainers: This category will bridge the gap between technologists and business leaders. Companies that deploy advanced AI systems will need a cadre of employees who can explain the

[8] Paul Daugherty and H. James Wilson, 'Process Reimagined', Accenture, 2018, available at https://www.accenture.com/_acnmedia/PDF-76/Accenture-Process-Reimagined.pdf#zoom=50, last accessed on 18 February 2020.

inner workings of complex algorithms to non-technical professionals, especially to overcome the 'black box' nature of algorithms.[9] Techniques such as Local Interpretable Model–Agnostic Explanations (LIME), which explains the underlying rationale and trustworthiness of a machine prediction, can be used by forensic analysts or explainers.

- Sustainers. This category of jobs will require workers to ensure that AI systems are operating as designed and that unintended consequences are dealt with required urgency. The role of ethics compliance managers or an ombudsman would become critical to upholding norms of human values and morals.

Reskilling and retraining would be the key to managing the transition while companies adopt AI technologies at varying speed. But the reskilling or training cannot be confined to a small group of people. It needs to be company-wide to develop a data-driven culture and mindset. One of the ways of upskilling the existing workforce is to create technology-infused

[9] In the process of their creation, machine-learning algorithms become so complex that they become unreadable except by their inputs and outputs. This is known as the 'black box' nature of algorithms as it becomes extremely difficult to explain the process of decision-making in a way that the average person can understand.

platforms and new tools that allow individuals to learn new skills. Kahoot, a game-based learning platform, is an example. Others include Coursera, edX, udemy, udacity, and Future Learn.

Adoption

Which Sectors Are Adopting AI Technologies?

The nature of jobs is also changing fast. As per LinkedIn data, between 2015 and 2017 AI skill penetration has risen in a wide range of sectors, with the maximum being in software and IT services, education, hardware and networking, and finance. Public administration, real estate, transportation and logistics appear to be have had the slowest rise in penetration.

Although the LinkedIn report cautions that changes driven by AI technologies may still be in their infancy, its impact across global markets cannot be dismissed. But it also states that the industries whose workforce has the most AI skills are also the ones that are changing or innovating the most. According to its data, while software and IT services continue to be changing the fastest, other sectors such as hardware and networking, education, finance, manufacturing, health care, construction, and consumer goods are fast catching up.

Experts predict that AI will have applications across nearly every sector from education and health care to construction, retail, and financial services. A look at the year on year growth of AI skills across various sectors gives a sense of the universality of AI in the future.

There are differing views about AI adoption among Indian industries. A Mckinsey study suggests that the potential for adoption of AI technology and services in Indian industries is somewhat low because of lower automation potential of activities and digital absorption; however its human capital, innovation foundation, connectedness, and labour market structure is within the global average.[10] According to a survey by EY, most of the exponential technologies (AI, IoT, robotics, big data, energy storage, 3D printing) are at a nascent stage of experimentation in the customer adoption cycle in India. Factors other than consumer pressure that may drive adoption by Indian companies include the rate of

[10] McKinsey Global Institute, 'Notes from the Frontier: Modeling the Impact of AI on the World Economy', September 2018, available at https://www.mckinsey.com/~/media/McKinsey/Featured%20Insights/Artificial%20Intelligence/Notes%20from%20the%20frontier%20Modeling%20the%20impact%20of%20AI%20on%20the%20world%20economy/MGI-Notes-from-the-AI-frontier-Modeling-the-impact-of-AI-on-the-world-economy-September-2018.ashx, last accessed on 18 February 2020.

falling cost curves, exports, presence of start-ups, and government regulation.[11]

On the other hand, a survey of companies conducted by the WEF projects a more optimistic picture in terms of adoption of AI technology by 2022 in India. In almost each of the technologies considered, from big data analytics to 3D printing to various kinds of robotics, India is above the global average in terms of proportion of companies likely to adopt them by 2022. Another survey, conducted by the Boston Consulting Group (BCG) in 2016, found that companies in the USA, China, and India have taken an impressive lead in adoption of AI over their counterparts in Japan, France, and Germany.[12] Transportation and logistics, banking, financial services, and insurance (BFSI), and automotive and technology companies are at the forefront of AI adoption.

AI across Industries: A Few Indicative Trends

Expectation from AI is running high but whether those expectations will be fulfilled depends on a

---

[11] NASSCOM, FICCI, and EY, 'Future of Jobs in India'.

[12] BCG, 'AI in the Factory of the Future: The Ghost in the Machine', 18 April 2018, available at https://www.bcg.com/en-in/publications/2018/artificial-intelligence-factory-future.aspx, last accessed on 18 February 2020.

variety of factors beyond just the mastery of data. Companies also need flexibility in management and organizational practices, require a AI strategy to be in place, and develop an intuitive understanding of AI. According to a global survey by MIT Sloan and BCG, only about one in five companies has incorporated AI in some offerings or processes. Only one in twenty companies has extensively incorporated AI in offerings or processes. Less than 39 per cent of all companies have an AI strategy in place. The largest companies—those with at least 100,000 employees—are most likely to have an AI strategy, but only half of such companies presently have one.[13]

According to another study by Tata Consultancy Services (TCS),[14] the most frequent user of AI is the IT department but the biggest beneficiaries of AI are supposed to be outside IT. Other users include

[13] BCG, 'Reshaping Business with AI', 2017, available at https://www.bcg.com/Images/Reshaping%20Business%20 with%20Artificial%20Intelligence_tcm9-177882.pdf, last accessed on 18 February 2020.

[14] 'Getting Smarter by the Sector: How 13 Global Industries Use Artificial Intelligence', TCS Global Trend Study: Part II, TATA, 2015, available at http://sites.tcs.com/ artificial-intelligence/wp-content/uploads/TCS-GTS-how-13-global-industries-use-artificial-intelligence.pdf, last accessed on 18 February 2020.

customer service, sales and marketing, finance, R&D, manufacturing, and HR.

AI by Domain: Global Scenario

The AI adoption and impact realities differ widely across sectors. In this section we will look at a few notable trends and practices.

In IT/ITES, AI technologies automate existing well-defined activities such as system administration, IT administration, business operations, and verification, and can create opportunities for new breakthrough kinds of activities that did not exist. Big tech companies such as Google, Amazon, Salesforce, Oracle, and Microsoft have improved their enterprise AI offerings helping companies integrate machine learning into their products. Google released Cloud AutoML where customers can bring their own data to train the algorithms to suit their specific needs. Amazon's 'AI as a Service' with Amazon AI aims to serve big and small time developers who want AI without the upfront costs or hassle. It unveiled offerings that will work like an API and allow any developer to access Lex (the NLP inside Alexa), Amazon Polly (speech synthesis), Amazon Rekognition (image analysis) and added video recognition, audio transcription, and sentiment analysis to its portfolio of services.

AI in fintech and BFSI prioritizes quick, personalized, and customized financial services. For example, Sun Life created and deployed a virtual assistant, Ella, to help users for Benefits and Pension by allowing them to stay on top of their insurance plans. Based on user data, the assistant sends users reminders such as 'Wellness benefits about to expire' or 'Your child will be off benefits soon'. AI also helps evaluate lenders and debtors speed up financial service processes, improve the customer experience, and can minimize errors. Chat bots are being used in banking to focus on search tasks. An example of this is Bank of America's bot called Erica that acts as a digital financial assistant for the bank's client base. Banks and insurance companies can use AI in profiling clients based on their risk score. Classification models such as XGBoost and ANN are trained on historical and pre-labelling data. Other uses of AI include underwriting services, automated claims process, churn prediction, contract analysis, valuation models, and algorithmic trading.

The application of AI in the transportation field is aimed at overcoming the challenges of increasing travel demand, CO_2 emissions, safety concerns, and environmental degradation. Examples of AI methods being used include ANNs, genetic algorithms, artificial immune system, Ant Colony Optimiser, Bee Colony Optimization, and Fuzzy Logic Model.

While cars currently have a number of AI-powered functionalities, such as anti-lock braking systems, airbag control, traction control systems, and electronic stability control, autonomous cars are the next frontier. Self-driving/autonomous vehicles have become a reality in the sea and sky since the 2000s. By 2015, we had Google's autonomous vehicles and Tesla's semi-autonomous cars. Experts predict that soon there will be self-driving and remotely controlled delivery vehicles, flying vehicles, and trucks. Peer-to-peer transportation services such as ridesharing are also likely to utilize self-driving vehicles. Beyond self-driving cars, advances in robotics will facilitate the creation and adoption of other types of autonomous vehicles, including robots and drones.

AI can address issues such as safety, reliability, efficiency, and pollution. Other than autonomous vehicles, AI can be used in traffic management and decision-making systems in order to enhance and streamline traffic management and make our roads smarter. Chinese e-commerce giant Alibaba has launched its traffic management service 'City Brain' in Kuala Lumpur, Malaysia. City Brain sorts through a mass of incoming data from 300 traffic lights, 500 CCTV cameras, public transport systems, and other streams in order to minimize road congestion. In Bengaluru, India, which regularly faces long traffic jams and the average speed on some roads at peak hours is just 4km/h (2.5mph), Siemens Mobility has built a prototype monitoring system that uses AI through traffic cameras. Traffic cameras automatically detect vehicles and this information is sent back to a central control centre where algorithms estimate the density of traffic on the road. The system then alters the traffic lights based on real-time road congestions.

Manufacturing tasks such as visual inspection, predictive maintenance, and even assembly can use AI. For example, a robotic prototype from Siemens automatically reads and follows CAD instructions to build parts without programming. Another company, Landing.AI, aims to help manufacturers incorporate AI into their workflows. For visual inspection, Landing. AI's system recognizes patterns of imperfections after 'viewing' only five product images. Manufacturers in many industries have long used robots to tackle complex assignments, but robots are evolving for even greater utility. For example, Kuka, a European manufacturer of robotic equipment, offers autonomous robots that interact with one another and with humans. Similarly, industrial robot supplier ABB launched a two-armed robot called YuMi that is specifically designed to assemble products (such as consumer electronics) alongside humans.

AI has found several areas of use in retail services as well. Chat bots and self-check-out services in retail are likely to benefit retailers and enhance customer satisfaction. H&M uses AI to analyse store returns, receipts, and loyalty cards to predict future demand for apparel and accessories and manage inventory. eBay uses AI-powered pricing and inventory algorithms to define the most appropriate prices for goods and notify sellers. Big brands such as Costco, Kohl's, Target, Tesco, and Walmart use either Google or Amazon AI technology along with smart devices to serve customers with easy and fast search. Me-Ality, a Canada-based tech start-up, has developed a virtual fitting kiosk that can scan a shopper's whole body. Alibaba has used AI in China to launch its first cashier- and cash-free wine store, Tao Cafe, as well as Futuremart.

How Are Indian Industries Applying AI?

Indian companies are also embracing the change by adopting AI in several processes across sectors. Telecom, media, and technology companies have been at the forefront of AI adoption. Indian IT companies have automated many of the routine tasks that were earlier performed by humans. For instance, Wipro is using its own cognitive platform called HOLMES, to automate multiple processes of its IT projects.

Indian banks have invested heavily in AI to boost their offerings and stay ahead in the tech adoption curve.

AI has allowed banks to develop a genuine relationship with customers over time as they have ventured into avenues such virtual assistants to understand customer queries.

Finally, Indian retailers, like Shoppers Stop, Reliance Retail and Aditya Birla Retail are also experimenting with AI, robotics, and cognitive technologies to build a seamless interface for customers.

The AI start-up ecosystem in India has included a few truly innovative experiments. For instance, GreyOrange designs and develops warehouse automation and technology solutions and offers products like Butler, a fleet of mobile robots for moving materials in warehouses more efficiently; Sorter, a fully automated sortation system to sort and divert outbound packets; and GreyMatter, a software platform for end-to-end intelligent order fulfilment.

Similarly, NetraDyne is a machine learning and deep learning company that focuses on computer vision and its applications to automotive and unmanned aerial systems navigation and collision avoidance. It also works on automated analysis of visual data collected by drones for verticals ranging from agriculture to site inspections.

Perfint Healthcare, a medical device technology company developing diagnostic equipment for the oncology space, has developed products such as Robio

EX (CT and PET-CT guided robotic positioning system), Robio EZ (robotic, mobile stand-alone system with 5 DOF for needle placement during CT Scan), and Maxio (image-guided, physician-controlled stereotactic accessory device to a CT system).

AI is also being used in diagnosing mental illness and age-related memory loss illnesses too. BrainSight is a software as a service (SaaS) product that detects major psychotic disorders. The software uses MRI scans to track brain activity over several minutes and records the visual. Currently, it can read six critical disorders including schizophrenia, bipolar disorder, Alzheimer's dementia, frontotemporal dementia, and schizoaffective disorder.

Gobasco, an agri-tech start-up, solves complex supply chain optimization problems. It identifies and treats the sources of inefficiencies with AI-powered processes and pipelines and cuts down the cost of supply chain and enhances commodity pricing. Other similar companies include Cropin, EM3 Agri Services, and Airwood.

The Overall Picture: Challenges and Opportunities

AI and cognitive solutions will not only change business process but entire business models that companies currently follow. The good news for India is it is leading this revolution from the front and is going to be one of the fastest adopters of AI-based services. The Indian government has also woken up to the potential of AI and has started putting in place a strategy to scale up and allocate resources for research and training. However, there are many challenges that India would have to overcome before it can become the 'AI garage' for emerging economies. These include the lack of big data, digital infrastructure, and highly trained manpower.

Globally as well how companies and countries choose to embrace AI will likely impact the outcomes and widen performance gaps between countries. Those that establish themselves as AI leaders (mostly developed economies) could capture an additional 20 to 25 per cent in economic benefits compared with today, while emerging economies may capture only half their upside. There could also be a widening gap between companies, with frontrunners potentially doubling their returns by 2030 and companies that delay adoption falling behind. For individual workers, too, demand—and wages—may grow for those with digital and cognitive skills and with expertise in tasks that are hard to automate, but shrink for workers performing repetitive tasks. There are also concerns regarding data privacy, biases in algorithms, and the regulatory ecosystem needed for AI to have a beneficial effect on society. The next chapter will deal with a few of these issues.

4

AI and Government in India

Overview

The disruptive potential of AI tools and techniques in business and the global economy is evident from the fact that it is referred to as the Fourth Industrial Revolution. While its impact on industry is much discussed, its potential in transforming governance is equally impressive. Here it holds the key to changing millions of lives through dramatically improved delivery of public services and unprecedented efficiency in the design of law and order and regulatory monitoring systems. In the hands of wise governments, AI can be a transformative tool. At the same time, AI abused by the powerful can easily lead, at least in imagination, to Orwellian states and dystopian political systems.

Governments, however, also have a critical role in not just harnessing AI wisely but also in developing and regulating it to prepare society for adopting it gradually. Given the transformative potential of AI at the workplace, governments have the crucial responsibilities of ensuring that AI applications create value for society, mitigating the adverse effects of job losses through safety nets and skill development, and protecting citizens from misuse of data. Failure on the part of policymakers to predict changes in society wrought by unhindered application of AI by private entities, especially in the jobs landscape, could lead to political backlash.

Countries round the world have started waking up to this reality and between 2017 and 2018, over 18 countries have devised national strategies to harness AI. This includes India as well.

In this chapter, we look at a few of the ways in which AI can potentially change public policy and governance and the AI applications around the world in this domain. We also explore the role of governments in development of AI and the policy questions AI throws up, with a particular focus on India's national strategy towards technological advancement in AI and related areas.

AI in Policy: A Promising Start

AI has the potential to solve many of the key and intractable challenges of public policy, particularly in development and public welfare areas, especially for a country as populous and diverse as India. Health care, for instance, is a key area of public service. India suffers from an acute shortage of doctors and health professionals as well as health facilities. Application of AI in health care can help address issues of high barriers to access to health care facilities, particularly in rural areas that suffer from poor connectivity and limited supply of health care professionals. AI-driven diagnostics, tracking and diagnosing the data captured by health workers, patient monitoring, early identification of potential pandemics, and imaging diagnostics, among others, can make a vast difference to both access to and cost of health care for the average Indian.

These are not just abstract possibilities. In a country with 60 million cardiovascular patients and less than 10,000 cardiologists, an AI start-up, CardioTrack, is today providing timely cardiac care, both diagnostic services and predictive medicines, in several tier-II cities. It helps bring portability to health care diagnostics by connecting AI to primary health care centres. CardioTrack sensors provide clinical grade reading for ECG, SpO2, and blood pressure.

Elsewhere Mumbai-based Qure.ai, established in 2016 and operating in multiple countries, is developing deep learning algorithms that interpret radiology images. It can read chest X-rays and CT scans and identify common abnormalities or scans with emergency findings. Bengaluru-based NeuroSynaptic Communications Pvt. Ltd aims to make health care more accessible and affordable by providing high-quality ReMeDi—Remote Healthcare Delivery Solutions, an app that collects information on various physiological aspects of patients remotely and provides them with its diagnosis.

Agriculture is not a field that one often associates with high-tech developments. Yet, AI holds the promise of driving a food revolution and helping meet the increasing demand for food. It also has the potential to address challenges such as inadequate demand prediction, lack of assured irrigation, and overuse/misuse of pesticides and fertilizers. Some use cases include improvement in crop yield through real time advisory, advanced detection of pest attacks, and prediction of crop prices to inform sowing practices. Bengaluru-based start-up CropIn uses AI to maximize per-acre value in agriculture. With its 'smartfarm' solution it is possible to geo-tag plots of farmland to find the actual plot area. It also helps in remote sensing and weather advisory, scheduling and monitoring farm activities for complete traceability, educating farmers on adoption of the right package of practices and inputs, monitoring crop health and harvest estimation, and alerts on pest, diseases, and such others. Tech giant

Microsoft's India operations has collaborated with ICRISAT, to develop an AI-based sowing app that uses machine learning and business intelligence from the Microsoft Cortana Intelligence Suite. The app sends sowing advisories to farmers on the optimal date to sow as it is one of the biggest challenges due to unpredictable weather conditions. Tata Rallis uses AI-powered drones to administer pesticides by harnessing data on crop health and soil conditions to increase output.

Education and skilling have remained difficult sectors for Indian policymakers. AI can potentially provide solutions to the quality and access issues observed in the Indian education sector. By some estimates, there are over 3,000 ed-tech start-ups in India. Potential use cases include intelligent tutoring systems (ITS), learning analytics, and automating and expediting administrative tasks. Delhi-based start-up Leverage Edu, an AI-enabled marketplace to help students with planning for higher studies, provides mentorship products, end-to-end college admissions guidance, programmes to help get first-job ready, as well as one to one virtual career advisory for multiple career streams. Intelligent tutoring system Sherlock can teach air force technicians to diagnose electrical systems problem in aircrafts.

Smart Mobility is beginning to change the face of transportation around the world and in ways very different from driverless cars. Potential use cases in this domain include autonomous fleets for ride sharing, semi-autonomous features such as driver assist, and predictive engine monitoring and maintenance. Other areas that AI can impact include autonomous trucking and delivery, and improved traffic management. Siemens Corporate Technology has developed an AI-powered smart traffic management system that is undergoing testing in Bengaluru. Delhi also plans to introduce high resolution cameras with sensor-based real time traffic volume count technology in 2019. Bengaluru-based Uncanny Vision has created an automatic number plate recognition solution that sends the number plate info over a secure network interface to the Toll Management Software with far improved recognition rates. It leads to faster flow of traffic, and is expected to cut toll plaza time by 50 per cent.

Environmental pollution—be it air, water, noise—and renewable energy are massive related challenges in India. Solid waste disposal and effluent treatment are other areas that need urgent attention. AI can help address many problems here through intelligent automation to estimate and control pollution, identification of critical pollutants, and prediction of meteorological events. Founded in California but headquartered in India, Gram Power provides cutting edge smart grid technology to address the growing electricity concerns in the country. The Central Water Commission of India is collaborating with Google to improve flood prediction systems through AI, geo-spatial mapping and hydrological data.

Public services offered through municipalities and panchayats can be vastly improved if AI technology is deployed strategically. Tech Mahindra has created a social media management product that collects grievances in a single platform from various channels and processes it. This solution resulted in 20–30 per cent improvement in citizen response times. Bengaluru-based VuNet Systems has developed a multi vector analytics platform that uses unsupervised machine learning (ML) techniques to improve Aadhaar-enabled payment systems.

Security, especially crime detection and prevention, can be revolutionized with AI technology. Some recent applications of AI are in the area of fingerprint analysis, recreating a face from the skull, creating images from pieces, and modern forensic methods. The police in Punjab and Uttar Pradesh are using facial recognition systems with options such as face search, text search, and so on. Punjab AI System (PAIS) has a database with more than 100,000 records of criminals housed in jails across the state. Trinetra, a product of Gurgaon-based start-up Aqu that Uttar Pradesh police is now using, has a database of approximately 500,000 criminals.

The defence department has also been taking tentative steps at deploying AI. The Defence Research and Development Organisation successfully tested the Rustom 2 drone (UAV) in February 2018 and is engaged in developing a 'Multi Agent Robotics Framework' (MARF), a system aimed at enabling collaboration among the various battlefield robots of the Indian army on surveillance and reconnaissance. It is also developing chemical, biological, radiological, nuclear, and explosive (CBRNe) UAVs to detect radiation, as well as remotely operated vehicles (ROVs) for surveillance and IED disposal. Among the paramilitary forces, the Border Security Force (BSF) is working on the Comprehensive Integrated Border Management System (CIBMS), developing an electronic surveillance system monitored by BSF personnel.

All these are among focus areas of the government for AI applications. The NITI Aayog is partnering with tech giants such as Google, Microsoft, IBM, ABB, and Intel to leverage their know-how in order to address some of India's public policy challenges. For instance, Google plans to work with the NITI Aayog on a range of initiatives including training and incubating Indian start-ups focused on AI and upskilling of Indian developers through machine learning crash courses.

IBM Watson has partnered with the government to develop AI models that could create smart cities. ABB will support the Make In India campaign through advanced manufacturing technologies that incorporate the latest developments in robotics and AI. In 2018, SAP will adopt 100 Atal Tinkering Laboratories for five years to nurture STEM learning among secondary school children across India. More broadly, companies are rapidly developing governance–enhancing AI solutions. Delhi-based start-up CivilCop, for instance, leverages its AI/ML and NLP solutions to empower citizens and governments by making grievance reporting and redressal faster and more efficient.

There are, of course, plenty of challenges as well. Lack of broad-based expertise in research and application of AI; absence of enabling data ecosystems like access to intelligent data; high resource cost and low awareness for adoption of AI; privacy and security, including a lack of formal regulations around anonymization of data; and absence of collaborative approach to adoption and application of AI all slow down the adoption of AI in various parts of governance and public service delivery.

But efforts persist. The Task Force on AI in the Commerce Ministry has a long list of focus areas for AI application. It includes manufacturing, fintech, health care, agriculture/food processing, education, retail/customer engagement, aid for differently abled, environment, national security, and public utility services. The broad approach of AI application in India's strategy documents revolves around the 'AI for All' concept. This means that India sees a disruptive technology like AI primarily as an opportunity to transform the quality of delivery of public service to ensure access to the most underserved population.

Developing AI: The Role of Government in Research, Funding, and Strategizing

If AI has the potential to transform governance, its development worldwide has also depended on critical government support. Research in AI across the globe has a long history of public funding with periodic ups and downs. However, the trend has moved towards private sector funding in the last two decades.

The BRAIN Initiative, created in 2013, is a 10-year, multibillion dollar fund for AI research in the USA, while the European Union's Human Brain Project envisages spending EUR 1 billion on AI over the next decade.[1] Other countries such as South Korea have

---

[1] Shashi Shekhar Vempati, 'India and the Artificial Intelligence Revolution', Carnegie India, 11 August 2016, available at https://carnegieindia.org/2016/08/11/india-and-artificial-intelligence-revolution-pub-64299, last accessed on 18 February 2020.

turned to the public–private model for funding AI research.

How has research in AI fared in India? Not surprisingly, it lags significantly behind the USA and Europe and more recently China and South Korea. According to a 2012 report by Professor Deepak Khemani of IIT Madras, AI research in India has been limited to a handful of passionate researchers with a focus on only certain areas such as machine translation, natural language, and text- and speech-related applications.[2] Unlike the USA, where significant research on AI is undertaken by DARPA, Indian AI research in defence is relatively limited. It is housed under the Centre for Artificial Intelligence and Robotics (CAIR), which is part of the DRDO. CAIR was established in 1986 and has worked on building integrated, networked, information systems, data mining tools, robotics, and other AI-enabled products for the Indian military. It has had some successes but nowhere near the scale of other players.

A more recent 2018 study (the Itihaasa report) shows that AI research has, however, progressed to areas like unsupervised learning, reinforcement learning, explainable AI, causal modelling, and blockchain.

[2] Deepak Khemani, 'A Perspective on AI Research in India', *AI Magazine* 33, no. 1 (Spring 2012): 96–8.

The report also cited that researchers seem to receive adequate funding support from government, industry, and universities. However, the report did identify challenges such as quality and quantity of students entering AI/ML research in India (the number of principal researchers in the country is in double digits), computing infrastructure, resources and administrative bottlenecks, lack of good-quality labelled data sets, and siloed research approach within universities.[3]

As we noted in Chapter 2, post 2013 AI started moving from the laboratories to commercialization. AI-powered start-ups started attracting venture capital funding. Industry-funded research projects witnessed a boost. The IITs and the IIITs have been at the forefront of this activity. The Pratiksha Trust has provided INR 300 million to IIT Madras to fund three chairs in the Centre for Computational Brain Research. At IIIT Hyderabad, Intel funded a project that led to the development of a dataset of Indian driving conditions. IIT Bombay has plans to collaborate with IBM in its AI Horizons Network as part of a multi-year collaboration to advance AI research. IIT Kharagpur plans to set up

---

[3] Itihaasa Research and Digital, 'Landscape of Artificial Intelligence/Machine Learning Research in India', 2018, available at http://www.itihaasa.com/pdf/Report_Final_ES.pdf, last accessed on 18 February 2020.

a Centre for AI, with seed funding from Capillary Technologies. At IIIT Delhi the Infosys Centre for AI was set up with a grant of INR 240 million from Infosys Foundation in 2016.

The government has also started strategizing about how to harness AI for India. In 2015, it approved the National Computing Mission at a cost of INR 45 billion and a seven-year period for implementation. One of the application areas of these supercomputers is AI. Within a span of a few months in 2018, three strategy reports on the direction that India should take on AI technology were published and committees set up.

An 18-member task force on AI led by Professor Kamakoti Veezhinathan of IIT Madras was set up on 24 August 2017 under the aegis of the Ministry of Commerce. Its purpose was to explore areas where AI can be leveraged for economic transformation, create policy and legal framework to accelerate deployment of AI technologies, and make recommendations for research programmes. The report, presented on 19 January 2018, recommends budgetary support for setting up of an inter-ministerial National AI Mission (N-AIM) with a budget of INR 2.4 billion per year for five years. Out of this budget, the task force proposes setting aside INR 500 million per year for core activities in AI research. It also identifies areas such as manufacturing, fintech, health care, agriculture, education, retail, national security, and environment where AI can add value.[4]

In February 2018, the Department of Defence Production, Ministry of Defence, set up a task force to study the use of artificial intelligence in defence as it has the potential to have a transformative impact on national security. The 17-member task force, led by N. Chandrasekaran, chairman, Tata Sons, submitted its report on 30 June 2018. It made recommendations about how to make India's defence sector a significant power in AI and the institutional interventions required to regulate and encourage a robust AI-based defence sector.[5]

On 8 February 2018, the Committee of Secretaries tasked the NITI Aayog with producing a national strategy plan for AI in consultation with ministries, academia, and industry. The NITI Aayog produced a discussion paper titled 'National Strategy for Artificial Intelligence: #AIForAll' in June 2018 with a view to

---

[4] Ministry of Commerce and Industry, 'Report of Task Force on Artificial Intelligence', March 2018, available at https://dipp.gov.in/whats-new/report-task-force-artificial-intelligence, last accessed on 18 February 2020.

[5] PIB, 'AI Task Force Hands Over Final Report to RM', PIB, 30 June 30 2018, available at https://pib.gov.in/newsite/PrintRelease.aspx?relid=180322, last accessed on 18 February 2020.

guiding the research and development in new and emerging technologies.[6]

In mid–2018, the Ministry of Electronics and Information Technology (MEITY) constituted four committees to create a policy framework and to develop the AI ecosystem in mid–2018.[7] These were the committee on platforms and data for AI (chairperson: Professor P.P. Chakraborty, IIT Kharagpur); the committee on leveraging AI for identifying national missions in key sectors (chairperson: Professor Rajeev Sangal, IIT-BHU); the committee on mapping technological capabilities, key policy enablers, skilling, reskilling and R&D (chairperson: Debjani Ghosh,

---

[6] NITI Aayog, 'National Strategy for Artificial Intelligence: #AIForAll', discussion paper, June 2018, available at https://niti.gov.in/writereaddata/files/document_publication/NationalStrategy-for-AI-Discussion-Paper.pdf, last accessed on 18 February 2020.

[7] PIB, 'Finalisation of National Artificial Intelligence Mission', 25 July 2018, available at https://pib.gov.in/newsite/PrintRelease.aspx?relid=181007, last accessed on 18 February 2020; *Economic Times*, 'AI Reports of MEITY Panels by Month-End', 6 April 2018, available at https://economictimes.indiatimes.com/small-biz/startups/newsbuzz/artificial-intelligence-reports-of-meity-panels-by-month-end/articleshow/63637110.cms?from=mdr, last accessed on 18 February 2020.

president, NASSCOM); and the committee on cyber-security, safety, legal, and ethical issues (chairperson: Professor Rajat Moona, director, IIT Bhilai).

In early 2019, the government announced a plan to launch an inter-ministerial National AI Mission (as recommended by the report of the task force on AI) and the establishment of a Centre of AI with support from six centres of excellence. However, no budgetary allocation has been made for this yet.

An Overview of National AI Strategies

As the race for becoming the global leader in AI heated up, many countries released strategies to promote the use and development of AI, including India. Of the 18 AI strategies released to date, 9 are fully funded and outline specific policies, while the other 9 are guiding documents that present objectives to guide future policymaking. There are also countries such as the USA and Israel who are global leaders in AI but have not yet articulated an explicit strategy.

Although the USA is still the undisputed leader in AI R&D and investments (in 2017 alone the unclassified budget of Pentagon shows that it spent USD 7.4 billion on R&D in AI and related fields; the military spends on R&D are huge but the budget is classified), it does not have a coordinated national strategy. The Obama administration made a start when it published three reports that laid the foundation of a US strategy: 'Preparing for the Future of Artificial Intelligence: Focus on Regulations, Ethics, Security'; 'National AI R&D Strategic Plan: Focus on Public Funded R&D'; and 'AI, Automation and Economy: Focus on Impact of Automation and Policies to Mitigate the Effects'.

The Trump administration has taken a more free market approach by focusing on removing regulatory barriers to innovation. It aims to maintain American leadership in AI, support the American worker and promote public R&D as declared at a summit on AI in May 2018. In 2018, the Department of Defense published an AI strategy paper 'Harnessing AI to Advance our Security and Prosperity'.

Israel has an excellent innovation ecosystem that benefits AI start-ups. However, the government does not yet have a targeted AI strategy in place. In June 2018, it announced the formation of a committee and around 10 to 15 professional subcommittees to devise the goals and plan of action for state policy on AI.

What Is India's AI Strategy and How Does It Compare with Other Countries?

As noted in the preceding sections, current AI strategies can be divided into two broad groups. The first group

Table 4.1A Funded Strategies *vs* Guiding Documents of Select Countries: Funded Strategies

Country	Release Date	Name of Report	Official Strategy	Funding (in USD)
Canada	March 2017	Pan-Canadian AI Strategy	Leverage leadership in AI research; attract global talent	95 million
Taiwan	January 2018	AI Action Plan	Make Taiwan a hub for AI development, industries, and applications	1.18 billion
France	March 2018	Strategy for AI	Safeguard French and European strategy; make AI environment friendly	1.75 billion
UK	April 2018	Industrial Strategy: AI Sector Deal	Develop AI in diagnostic health; become leader in ethical AI	1.24 billion
South Korea	May 2018	Artificial Intelligence R&D Strategy	Lead global AI R&D investments	1.95 billion

(*Cont'd*)

Table 4.1A   (Cont'd)

Country	Release Date	Name of Report	Official Strategy	Funding (in USD)
Germany	November 2018	Federal government's AI Strategy: AI made in Germany	Reimagine work in the AI age; enriching work–life balance	3.5 billion (up to 2025)

Sources: Prepared by the authors based on Tim Dutton, 'Building an AI World: Report on National and Regional AI strategies', CIFAR, 6 December 2018 (available at https://www.cifar.ca/docs/default-source/ai-society/buildinganaiworld_eng.pdf?sfvrsn=fb18d129_4, last accessed on 13 April 2020); Observer Research Foundation, 'In Pursuit of Autonomy: AI and National Strategies', 2018 (available at https://www.orfonline.org/wp-content/uploads/2018/11/In-Pursuit-of-Autonomy-AI-and-National-Strategies.pdf, last accessed on 13 April 2020); German AI Strategy (available at https://www.de.digital/DIGITAL/Redaktion/EN/Standardartikel/artificial-intelligence-strategy.html, last accessed on 13 April 2020).

Table 4.1B Funded Strategies *vs* Guiding Documents of Select Countries: Guiding Documents

Country	Release Date	Guiding Document	Overall Focus	Implementation Status
Japan	March 2017	AI Technology Strategy	Build on success in robotics; reinvigorate productivity through AI	Yes
China	July 2017	A Next Generation AI Development Plan	Drive military–civil fusion, that is, knowledge sharing; establish global technical standards in AI	Yes
India	June 2018	National Strategy for AI: #AIforAll	Use AI for inclusive growth; archive AI relevant data and make it accessible	No

Sources: Prepared by the authors based on Tim Dutton, 'Building an AI World: Report on National and Regional AI strategies,', CIFAR, 6 December 2018 (available at https://www.cifar.ca/docs/default-source/ai-society/buildinganaiworld_eng.pdf?sfvrsn=fb18d129_4, last accessed on 13 April 2020); Observer Research Foundation, 'In Pursuit of Autonomy: AI and National Strategies', 2018 (available at https://www.orfonline.org/wp-content/uploads/2018/11/In-Pursuit-of-Autonomy-AI-and-National-Strategies.pdf, last accessed on 13 April 2020); Federal Ministry of economic Affairs and Energy, 'The Federal Government's Artificial Intelligence Strategy: "AI Made in Germany"' (available at https://www.de.digital/DIGITAL/Redaktion/EN/Standardartikel/artificial-intelligerce-strategy.html, last accessed on 13 April 2020).

comprises strategies that, when first announced, included specific policies and funding. Funding varies significantly ranging from less than USD 25 million (Australia) to USD 2 billion (South Korea). The second group is made up of strategies that feature 'guiding' documents. These strategies were not funded when first announced; instead, they outlined strategic objectives to guide future policymaking. Of these, China and Japan have already begun implementation while India lags behind.

Each AI strategy is unique and focuses on select aspects of AI policy. These strategies have been classified according to overall focus/strategy or according to the role the state takes. The CIFAR report categorizes them into four main types: research and talent; industrialization, comprehensive, and guiding. A London Business School report categorizes the policies according to the role of the state: enabler, driver, or interventionist.

The strategy reports touch upon the following multiple areas of public policy.

- **R&D:** The creation of new research centres, hubs, or programmes in basic and applied AI research or a commitment to increase existing funding for public AI research.
- **Skilling:** Initiatives to help students and the overall labour force develop skills for the future of work,

such as investments in STEM (science, technology, engineering, and mathematics) education, digital skills, or lifelong learning.

- **Academia:** Funding to attract, retain, and train domestic or international AI talent, including funding for chairs and fellowships or the creation of AI-specific masters and PhD programmes.

- **Industry:** Programmes to encourage private sector adoption of AI technologies, including investments in strategic sectors, funding for AI start-ups and small and medium-sized enterprises (SMEs), and strategies to create AI clusters or ecosystems.

- **Ethics:** The creation of a council, committee, or task force to create standards or regulations for the ethical use and development of AI. This area also includes specific funding for research or pilot programmes to create explainable and transparent AI.

- **Digital infrastructure:** Funding for open data partnerships, platforms, and datasets, as well as commitments to create test environments and regulatory sandboxes.

- **AI in government:** Pilot programmes that use AI to improve government efficiency, service delivery, and public administration.

- **Inclusion:** Ensuring that AI is used to promote social and inclusive growth and that the AI community is inclusive of diverse backgrounds and perspectives.

How Much Funding Are Governments Committing to AI?

It should hardly come as a surprise that the USA and China are the biggest spenders on AI ecosystem. The Chinese government is pumping in upward of USD 7 billion; some estimates forecast a USD 150 billion AI industry by 2030. The USA has no central AI policy, but individual projects are funded by military and paramilitary departments such as DARPA, IARPA, and Military Service Research Laboratories. It plans to make longer-term and stable funding available to attract the best academics to invest in research–relevant areas of defence. While India has made lofty strategies, the government has only allocated USD 480 million in funds for its 'Digital India' initiative. No government funding has yet been announced for the National AI programme although it has promised to invest in research, training, and skill development in robotics, AI, digital manufacturing, big data intelligence and quantum communications, among other initiatives. Commitments to AI in annual budget of the leading countries range from USD 5 billion (USA), USD 4 billion (China) and USD 720 million (Japan). Other countries are also catching up on providing funding for AI with committed funding for the next five years: USD 22 million in Australia, USD 90 million in the UK, USD 93 million in Canada, USD 110 in Singapore, USD 300 in Taiwan, USD 1700 million in France, and USD 2000 in South Korea.

Roadmap and Way Forward

The centrality of AI in shaping every country's social, economic, and political trajectory is borne out by the rapidity with which governments have evolved national strategies. Each aims to retain its dominance in existing industries or make new in-roads. Governments of developed and emerging economies are also focusing on minimizing the social disruptions that will be inevitable—loss of jobs in the short run and the need for reskilling workers. Lastly, governments are trying to harness AI to improve people's lives by solving some of the world's most vexing challenges and inefficiencies.

The Indian government has taken the first steps in such an endeavour. However, many challenges remain. These include the absence of collaborative efforts between various stakeholders; lack of enabling data ecosystem; low intensity of AI research; inadequate availability of AI expertise, manpower, and skilling opportunities; low awareness for adopting AI in business opportunities; unclear privacy, security, and ethical regulations; and unattractive intellectual property regime. These challenges would need to be addressed in an expeditious manner in order for India to be able to reach its AI goals as laid out in the strategy documents.

AI technology could also be used in other organs of the state, namely parliament and the judiciary. In fact, the Ministry of Parliamentary Affairs announced that the two houses will soon use AI and ML to streamline their legislative duties as part of the National E-Vidhan project. NeVa, a new flagship programme, aims to create a data depository by bringing together all the legislatures of the country on one platform and use AI and ML analysing the data. India could also move to e-Parliament to improve efficiency. The Indian Parliament, with about 800 members, could meet online by investing in a high-tech P2P website where each executive can log in securely, speak when it is his turn, even leave supplementary opinions and materials in a pre-allotted e-space for others to watch if needed. Some countries such as Korea have already taken the plunge.

The judiciary could benefit from AI too by minimizing procedural delays, ensuring predictability of the judicial process, and ensuring transparency of judges' work, but such processes cannot be limited to algorithms and must take into account particular circumstances and ensure respect for fundamental rights. AI has been playing a role in automating divorces, contract analysis, and conducting due diligence. The Indian judicial system could adopt AI to fix logjams, address transparency issues, and help with case prediction. Law firms such as Cyril Amarchand Mangaldas have partnered with Toronto-based Kira Systems to transition to a digitized environment.

Even as AI technology seems to be impacting every sector, with governments rushing to embrace it, there remains a number of questions about the regulatory framework that should govern AI. The risks related to privacy, security, and data protection need to be understood clearly by all stakeholders, namely citizens, government, and industry, before it becomes too late. The next chapter examines these issues in some detail.

5

Opportunities, Challenges, and Next Steps

Debating AI: The Good, the Bad, and the Ugly

In Douglas Adam's *Hitchhiker's Guide to the Galaxy* (1979), the superintelligent computer Deep Thought was built to answer the question, what is the meaning of 'life, the universe, and everything'. It then built the biggest and most complex computer in existence, planet Earth, to answer this vexing question. While answers to such existential questions remain as enigmatic as ever, the concept of artificial intelligence has moved from the realms of science fiction to reality. AI's disruptive effect is already visible in many products and services, from chat bots, facial recognition software, and digital assistants such as Siri, Alexa, and Cortana to medical diagnostics, meteorological predictions, and traffic management. The potential for AI cuts across industries and more companies are adopting the technology to improve business performance and enhance productivity and innovation.

Alongside the economic benefits and challenges, AI's impact on society is inevitable but its exact nature is unclear. Experts are divided in their opinions. On the one side are Facebook's Mark Zukerberg and Alphabet's former CEO Larry Page who are strong proponents of AI and do not foresee much danger to humanity. In fact, Zukerberg believes in deeply integrated AI to help fight terrorist propaganda on Facebook. On the other side are stalwarts such as SpaceX's Elon Musk, Microsoft's Bill Gates, founding engineer of Skype, Jann Tallin, and renowned physicist Stephen Hawking who see AI as a potential threat to humanity, rating it higher than a nuclear threat, and caution that instead of working on making AI more capable, research should focus on maximizing the societal benefit of AI and making it robust. As early as 2015, both Musk and Hawking, who sat on the scientific advisory board of Future of Life Institute, became signatories to an open letter drafted by the researchers at the institute. The letter, directed to the broader AI community, highlighted the pros and cons of AI and outlined a research agenda[1] focused on maximizing societal benefits. It said, 'Success in the quest for artificial intelligence has the potential to bring unprecedented benefits to humanity, and it is therefore worthwhile to investigate how to maximize these benefits while avoiding potential pitfalls.' To date, the letter has been signed by over 8,000 people, many of whom are domain experts.[2]

The research agenda outlined in the letter included the following: (*a*) optimizing AI's economic impact through labour market forecasting, policy for managing adverse effects, and economic measures; (*b*) research on law and ethics, especially how to regulate autonomous vehicles and autonomous weapons, privacy, and the kind of policies that are required; and (*c*) research on robust AI to ensure that the system performs as desired.

There is no doubt that AI could be harnessed for the benefit of humanity, from health care to climate change and humanitarian crises, as has been discussed

[1] Stuart Russell, Daniel Dewey, Max Tegmark, 'Research Priorities for Robust and Beneficial AI', *AI Magazine*, 2015, available at https://futureoflife.org/data/documents/research_priorities.pdf, last accessed on 18 February 2020.

[2] Future of Life Institute, 'An Open Letter: Research Priorities for Robust and Beneficial AI', 2015, available at https://futureoflife.org/ai-open-letter/, last accessed on 18 February 2020.

in the preceding chapters. But there are many risks and challenges that need to be considered and mitigated before embracing it wholly. Currently, despite the progress made, many difficult problems remain that will require more scientific breakthroughs. For example, most of the progress is in what is called 'narrow AI', where machine learning techniques are developed to solve specific problems. The harder to crack problem is what is referred to as 'artificial general intelligence', where the challenge is to develop AI that can tackle general problems in the same way that humans can. Furthermore, obtaining data sets that are sufficiently large and comprehensive to be used for training is often extremely difficult. In applications where trust matters and predictions carry societal implications, such as in criminal justice applications or financial lending, the 'black box' complexity of deep learning techniques creates the challenge of 'explainability' or showing which factors led to a decision or prediction and how.

Given that adoption of AI is quite uneven across industries and countries, there is time to develop possible regulatory frameworks keeping ethical considerations and risk mitigation in the forefront. It is heartening to see that most countries have taken a proactive step in developing a strategy for adoption of AI and ethics and regulations are part of almost all the strategies. The role of governments is crucial in ensuring that regulations go beyond acting as facilitators or enablers of AI. They need to soften the disruptive transitions that will accompany it, and place a new focus on ethical and responsible use by enforcing data privacy and cybersecurity, and by eliminating algorithmic biases. Tech giants such as Google, Microsoft, and Intel have also started grappling with many of the policy questions that arise with the use of AI. Since the intended and unintended consequences of adopting AI across sectors is as yet unclear, governments need to work with civil society, academia, and AI practitioners to frame policies and guidelines on responsible AI development and use, explainability standards, fairness appraisal, safety considerations, human–AI collaboration, liability frameworks, and many other areas.

In addition, the regulatory frameworks adopted by each country will depend on the trade-offs that the country is willing to make: should there be restrictions on levels of automation in labour-heavy industries or should the government provide safety nets and reskilling options to those displaced by AI? Should tools like drones or predictive policing be used by the military or law enforcement or do the risks of safety and algorithmic biases outweigh the advantages? To make these decisions, there needs to be clear understanding of the opportunities and challenges that AI-powered technologies offer.

In this chapter, we discuss some of the ethical challenges that AI brings to the forefront and the regulatory frameworks that need to be put in place to ensure that AI remains a force for good.

Promises, Challenges, and Threats of AI

As we enter the AI age, the promise of AI techniques is visible in speech recognition, language comprehension, vehicle navigation, data analytics, and predictive analytics. It has contributed to or shown potential for increased economic productivity, more accurate medical screening and diagnosis, more precise targeting of beneficiaries through reduction in errors, and minimized human fatalities in risky industries such as mining and in the military.

The promise is also hampered by a number of challenges that disrupt the deployment of AI on a larger scale. Some issues are specific to India and some are applicable globally.

Further, AI also poses a variety of threats. These could be direct or malicious, that is, individual or organization deployed or ones that compromise AI technology to undermine the security of an individual or organization. There could be indirect threats such as mass unemployment or unintentional threats such as algorithmic bias.

In this section, we discuss the challenges as well as threats that AI poses to humanity.

Challenges

Scarcity of big data: The most powerful AI machines are the ones that are trained for supervised learning. This training requires labelled data, that is, data that is organized to make it ingestible for machines to learn. However, the availability of well-labelled, feature-rich local data sets is extremely limited in India. A few government bodies make some data sets available but they are limited in number and scope. For instance, the Reserve Bank of India (RBI) maintains a database on the Indian economy; the Indian Space Research Organisation (ISRO) provides some datasets from its satellites via its mapping service Bhuvan; the Wildlife Institute of India provides some datasets that it tracks and maintains. Even the government's open data platform started in 2012 is sketchy. Critical datasets are not available on data.gov.in. Available datasets are often outdated, duplicated, incomplete, inadequately referenced, and lack common terms used to describe the data. Top-level metadata such as data collection methodology and a description of the variables are also either missing or incomplete. These shortcomings make it difficult to compare and analyse datasets properly. Organizations are trying to get around this issue by investing in design methodologies and trying to figure out how to make AI models learn despite the scarcity of labelled data. 'Transfer Learning', 'Unsupervised/Semi-Supervised Learning', 'Active Learning', and so on are just a few examples of the next-generation AI algorithms that can help resolve this.

Lack of clean data: For data to be used to train AI, it needs to be recorded in consistent, machine readable formats for accuracy and it has to be ensured that it does not present the algorithms with unintended biases. This is a particularly big problem in India as a lot of its data is not digitized or is in unstructured format. For example, India does not have a unified platform to access health care facilities at an affordable price. This basically means that there is no available data repository, which, if it was there, could be hugely beneficial for both medical practitioners and patients as it becomes easier to track past medical history and also to provide better solutions by leveraging technologies such as AI and ML. Some companies are using international datasets to overcome the problem. Chennai-based genomic intelligence company Kyvor Genomics uses AI models to develop its cancer therapy solution called CANLYTx. It is an AI-based system that involves a diagnostic test that identifies patients most likely to be helped or harmed by a new medication, and based on that analysis it zeroes in on targeted drug therapy. The company currently does not require local datasets as it is working with internationally available drugs for cancer treatment. Mohali-based agritech start-up Agnext developed a spectral analysis device that analyses the curcumin content in a particular turmeric harvest. Initially, they worked with worldwide crowd-sourced datasets but they have started collecting data from labs across the country to build datasets.

Data localization: The act of storing data on any device that is physically present within the borders of a specific country where the data was generated is known as data localization. Free flow of digital data, especially data which could impact government operations or operations in a region, is restricted by some governments for security concerns. However, some experts oppose the move as it is seen as hindering the flexibility of the internet and adding to the cost for global companies that have to maintain multiple local data centres. Last year, RBI in India issued a circular mandating that payment-related data collected by payment providers must be stored only in India, setting a 15 October 2018 deadline for compliance. This covered not only card payment services by Visa and MasterCard but also of companies such as Paytm, Whatsapp, and Google, which offer electronic or digital payment services. Countries such as China, Russia, and Brazil also have strong data localization laws, but China also has local datasets which can be used to train algorithms. India does not have that capability, and this could impact start-ups looking to attain global stature as reciprocal restrictions could be slapped by other countries. The USA, EU, and Australia on the other hand allow free flow of cross-border data to varying degrees.

Limited technical capacity: AI algorithms are usually very complex, often requiring thousands of calculations, sometimes even more, computed every second. With the development of cloud and distributed processing over the past decade, it has become possible to process big data algorithms, ushering in the current age of AI-powered data analytics. However, as demand for more powerful processors increases, bottlenecks will start emerging, making it difficult for enterprises to adopt the technology. For start-ups and small and medium businesses, this would mean raising huge sums of capital to bring on board better processors and larger storage servers, which many would struggle to do. This trend also means businesses will have a hard time securing data across multiple, non-relational databases that are constantly evolving.

Impact on jobs: The rapid advance in AI technology has sparked concerns about how it would impact employment. There is fear that as AI improves, it could supplant workers, creating a pool of unemployable humans who cannot compete economically with machines. While there is no definitive way to predict the scale of job losses or quantify the new jobs that will be created, various studies have attempted to address this question with differing results. For instance, as mentioned in Chapter 3, the study by Frey and Osborne predicted that some functions within 47 per cent of jobs will be automated. They ranked occupations by probability of computerization and predicted that jobs such social science research assistants, atmospheric and space scientists, and pharmacy aides had a 65 per cent or higher probability of automation. Another report on OECD countries applied a different 'whole occupations' methodology and put the share of jobs potentially lost to computerization at 9 per cent. The WEF's 2018 report, however, predicts that a net of 58 million new jobs would be created due to the disruption caused by AI. Most studies[3] consistently

[3] Mark Muro, Robert Maxim, and Jacob Whiton, 'Automation and Artificial Intelligence: How Machines Are Affecting People and Places', January 2019, available at https://www.brookings.edu/research/automation-and-artificial-intelligence-how-machines-affect-people-and-

predict that the least well-off will suffer the most from automation. But a new study by Brookings, published in 2019, gives a different prediction. While stating that almost all occupations can be impacted by AI, it shows through a comparative textual analysis of text of AI patents and the text of job descriptions that it would affect better-paid, white-collar occupations such as market research analysts, sales managers, computer programmers, and personal financial advisors more than low-paying, hands-on services such as personal care, food preparation, or health care.[4]

---

places/, last accessed on 18 February 2020; McKinsey Global Institute, 'A Future That Works: Automation, Employment and Productivity', 2017, available at https://www.mckinsey. com/~/media/mckinsey/featured%20insights/Digital%20 Disruption/Harnessing%20automation%20for%20a%20 future%20that%20works/MGI-A-future-that-works-Executive-summary.ashx, last accessed on 18 February 2020; Arntz, Gregory, and Zierahn, 'The Risk of Automation for Jobs in OECD Countries'.

[4] Mark Muro, Jacob Whiton and Robert Maxim, 'What Jobs Are Affected by AI? Better-Paid, Better-Educated Workers Face the Worst Exposure', Metropolitan Policy Program, Brookings, November 2019, available at https:// www.brookings.edu/research/what-jobs-are-affected-by-ai-better-paid-better-educated-workers-face-the-most-exposure/, last accessed on 18 February 2020.

Threats

A number of experts from various disciplines came together to contribute to a report that lays out the risks associated with AI being used with malicious intent. The report focuses on areas of AI that are available now or likely to be available within five years. Published in February 2018, the report warns that AI is ripe for exploitation by rogue states, criminals, and terrorists.[5]

There are several security-relevant properties of AI that make it vulnerable to abuse. It is important to bear in mind that AI is a dual use area of technology—in other words, it can be put towards both civilian and military use and, more broadly, towards beneficial or harmful ends. AI systems are commonly both efficient and scaleable: once trained, AI can complete tasks more quickly and cheaply than humans. It can also compute higher volume of data with higher computing power, and can far exceed human capabilities. AI systems can

---

[5] Miles Brundage et al., 'The Malicious Use of AI: Forecasting, Prevention and Mitigation', Future of Humanity Institute, University of Oxford, Centre for the Study of Existential Risk, University of Cambridge, Center for a New American Security, Electronic Frontier Foundation, and OpenAI, February 2018, available at https://arxiv.org/ftp/arxiv/papers/1802/1802.07228.pdf, last accessed on 18 February 2020.

increase anonymity and psychological distance. Finally, AI developments lend themselves to rapid diffusion: while attackers may find it costly to obtain or reproduce the hardware associated with AI systems, such as powerful computers or drones, it is generally much easier to gain access to software and relevant scientific findings.

Today's AI suffers from a number of novel unresolved vulnerabilities. These include data poisoning attacks (introducing training data that causes a learning system to make mistakes), adversarial examples (inputs designed to be misclassified by ML systems), and the exploitation of flaws in the design of autonomous systems' goals. They demonstrate that while AI systems can exceed human performance in many ways, they can also fail in ways that a human never would.

Given these features of AI, it can pose a threat to physical security as well as political security. Instances of enhanced physical threat posed by AI systems include terrorist repurposing of commercial AI systems: commercial systems can be used in harmful and unintended ways, such as using drones or autonomous vehicles to deliver explosives and cause crashes.

Another threat is of escalation—endowing low-skill individuals with previously high-skill attack capabilities. AI-enabled automation of high-skill capabilities such as self-aiming, long-range sniper rifles reduce the expertise required to execute certain kinds of attack.

AI can also help increase the scale of attacks. Humans and machines can team up to use autonomous systems, increasing the amount of damage that can be inflicted. An example is one person launching an attack with many weaponized autonomous drones.

Moreover, AI can enable swarming attacks. Distributed networks of autonomous robotic systems, cooperating at machine speed, provide ubiquitous surveillance to monitor large areas and groups and execute rapid coordinated attacks.

Finally, AI can move attacks further in time and space: physical attacks are further removed from the actor initiating the attack as a result of autonomous operation, including in environments where remote communication with the system is not possible.

Among the threats to political security the key one comes from the state. The state can use automated surveillance platforms to suppress dissent. A nation's surveillance powers can be extended by automating image and audio processing, permitting the collection, processing, and exploitation of intelligence information at massive scales for myriad purposes, including the suppression of debate. For instance, Indian researchers from the University of Cambridge, India's National Institute of Technology, and the Indian Institute of Science presented a paper on a deep learning techniques for facial image recognition to identify partially obscured faces. While the intended purpose is to nab criminals, it could be used by the state to target protesters and dissidents who conceal their faces at protests and demonstrations to protect their identity.

AI can also help mislead and confuse. This may occur through circulation of fake news reports with realistic fabricated videos and audio. Highly realistic videos showing inflammatory comments by influencers that they never actually made may also be created. Automated, hyper-personalized disinformation campaigns can be launched using AI. Individuals can be targeted in swing districts with personalized messages in order to affect their voting behaviour.

AI can be used to automate influence campaigns. For example, AI can analyse social networks to identify key influencers, who can then be approached with offers or targeted with disinformation. In denial-of-information attacks, bot-driven, large-scale information generation attacks are leveraged to swamp information channels with noise (false or merely distracting information), making it more difficult to acquire real information. Information availability may be manipulated by using the content curation algorithms of media platforms to drive users towards or away from certain content to manipulate user behaviour.

In addition to these threats which have malicious intent, there are threats that are unintentional or system

119

related. Take, for instance, algorithmic bias. Algorithmic bias occurs when a computer system reflects the implicit values of the humans who created it. While generally the blame for bias in AI is put on the training data, the reality is that bias can creep in long before the data is collected as well as at many other stages of the deep learning process: during the framing of the problem, collecting data, and preparing the data. For example, biases creep in during hiring decisions as Amazon found that its internal recruiting tool was dismissing female candidates because it was trained on historical hiring decisions that favoured males over females.[6]

The use of AI-based tools in law enforcement has thrown up a number of cases of algorithmic biases. For example, in the USA, a recidivism estimation algorithm, COMPAS, was used in the criminal justice system for parole hearings. The COMPAS system was shown to give systematically biased results. This bias, compounded by the misguided use of the system for bail and sentencing proceedings, led to significant

---

[6] Karen Hao, 'This Is How AI Bias Really Happens—and Why It's So Hard to Fix', *MIT Technology Review*, 4 February, 2019, available at https://www.technologyreview.com/s/612876/this-is-how-ai-bias-really-happensand-why-its-so-hard-to-fix/, last accessed on 18 February 2020.

inequities in criminal sentencing outcomes in the courts utilizing the technology.

In India, there is under-counting of crimes as the National Crime Records Bureau only records the 'principal offence' whenever a first information report (FIR) is filed, which could mean that in a scenario where there is both rape and murder, murder could remain uncounted. This has huge implications for the future of predictive policing.[7]

Towards a 'Responsible AI': Trends in India and the World

The need for ethics and laws to regulate AI is seen as critical for it to gain the confidence of the public. If AI leads to privacy violations, bias, or malicious use, or if much of the world comes to blame it for exacerbating inequality, the potential of AI would remain unfulfilled. Establishing confidence in its abilities to do good, and at the same time addressing misuses will be crucial.

[7] Arindrajit Basu and Elonnai Hickok, 'AI in the Governance Sector in India', The Center for Internet and Society, 2018, available at https://cis-india.org/internet-governance/ai-and-governance-case-study-pdf, last accessed on 18 February 2020.

This has prompted many countries to take proactive steps to frame policies to regulate AI. At the same time, tech giants such as Google, Intel, and Facebook have declared ethical standards they plan to adhere to.

India has also woken up to the need to regulate AI and has taken some small steps in that direction.

What Are the Steps Taken by India to Regulate AI?

Currently, there is no comprehensive law that governs AI in India. Over the years, the government has made piecemeal policies to protect certain facets related to AI, namely data privacy and data localization, but there is not much discussion on possible regulatory issues beyond these. In fact, NITI Aayog's AI Strategy Discussion Paper published in 2018 in its section on ethics makes recommendations mostly on the issue of data privacy. Box 5.1 presents a summary of its key points.

At present, the usage of personal data or information of citizens is regulated by the Information Technology (Reasonable Security Practices and Procedures and Sensitive Personal Data or Information) Rules, 2011, under Section 43A of the Information Technology Act, 2000. The rules define personal information of an individual as any information which may be used to

Box 5.1   Recommendations Related to Ethics and Privacy in NITI Aayog's Paper

- **Establish a data protection framework with legal backing:** This should be based on the seven core principles of data protection and privacy: informed consent, technology agnosticism, data controller accountability, data minimization, holistic application, deterrent penalties, and structured enforcement.
- **Establish sectoral regulatory frameworks:** In addition to a central privacy protection law, sectoral regulatory frameworks are needed to protect user privacy and security. Japan and Germany have developed new frameworks applicable to specific AI issues such as regulating next generation robots and self-driving cars respectively.
- **Benchmark national data protection and privacy laws with international standards:** The EU's General Data Protection Regulation (GDPR) guidelines, which have been enforced in May 2018, and the French laws could act as guidelines for India's privacy protection regime.
- **Encourage AI developers to adhere to international standards:** AI practitioners from across the world have acknowledged the need to frame standards for AI and many have set out guidelines to be followed. Indian enterprises need to step up too.

(Cont'd)

- **Encourage self-regulation:** Data Privacy Impact Assessment Tools can be used by AI developers and enterprises to manage privacy risks.
- **Invest and collaborate in privacy preserving AI research:** New mathematical models for preserving privacy are undergoing research and development and India should collaborate on areas of research such as differential privacy, privacy by design, safety-critical AI, and multi-party computations which enable protection of privacy despite data sharing on a wide scale.
- **Spread awareness:** The Supreme Court of India has termed privacy as a fundamental right but this right can be protected not only by laws enforced by the state but also by making citizens aware of their rights and how they can protect them.

Source: 'National Strategy for Artificial Intelligence: #AIForAll,' Discussion Paper, NITI Aayog, June 2018.

identify them. They hold the body corporate (who is using the data) liable for compensating the individual in case of any negligence in maintaining security standards while dealing with the data.

There were some attempts to frame data privacy laws between 2011 and 2015. A group of experts under Justice A.P. Shah had submitted a report on privacy in October 2012. The report proposed a framework for a Privacy Act in India which 'must include privacy-related concerns around data protection on the internet and challenges emerging therefrom'. A draft bill was also prepared but did not reach the Parliament.

The introduction of Aadhaar, India's first biometric identity card for residents, in 2012 brought the issue of data privacy to the forefront. The Aadhaar database captures biometric information (finger prints and retina scan) as well as basic demographic information (name, age, address, photo) and can be used to authenticate the identity of a person who wishes to avail a service provided by the government or a private sector organization. The concern with data privacy arises with the creation of a large database of residents and its use by third party service providers. Currently, the Aadhaar Act is silent on the powers of the UIDAI to take enforcement action against errant companies in the Aadhaar ecosystem. This includes companies wrongly insisting on Aadhaar numbers, those using Aadhaar numbers for unauthorized purposes, and those leaking Aadhaar numbers, all of which have seen several instances in the recent past. Each of these can affect informational privacy and requires urgent redressal.

In this context, the right to privacy of citizens is the pillar on which India's data protection regime has to be built. Since this right is not mentioned explicitly in the constitution, the matter has gone to court a number of times, the latest being in the *Justice K.S. Puttaswamy (Retd.)* v. *Union of India* case.[8] Previous judgments on the right to privacy were in the context of the right to property and the surveillance powers of the state.[9]

In 2012, a petition was filed in the Supreme Court, challenging the constitutional validity of Aadhaar on the grounds that it violated an individual's right to privacy. The matter got referred from a three-judge bench to a five-judge bench and eventually to a nine-judge bench of the Supreme Court for an authoritative pronouncement on the status of the right to privacy. The bench gave its ruling on 24 August 2017 in what is referred to as the Puttaswamy case and affirmed the constitutional right to privacy. It declared privacy to be an integral component of

[8] *Justice K.S. Puttaswamy (Retd.)* v. *Union of India*, 1 2017 (10) SCALE 1, available at https://indiankanoon.org/doc/127517806/.

[9] *M.P Sharma* v. *Satish Chandra* (1954) and *Kharak Singh* v. *State of Uttar Pradesh* (1962), available athttps://indiankanoon.org/doc/1306519/ and https://indiankanoon.org/doc/619152/ respectively.

Part III of the Constitution of India, which lays down the fundamental rights. These fundamental rights cannot be given or taken away by law, and all laws and executive actions must abide by them. The Court also observed that 'informational privacy', or the privacy of personal data and facts, is an essential facet of the right to privacy.[10]

The Supreme Court, however, clarified that like most other fundamental rights, the right to privacy is not an *absolute right*. Subject to the satisfaction of certain tests and benchmarks, a person's privacy interests can be overridden by competing state and individual interests. Thus, a violation of privacy in the context of an arbitrary state action would attract a *reasonableness* enquiry under Article 14. Similarly, privacy invasions that implicate Article 19 freedoms would have to fall under the specified restrictions under this constitutional provision like public order, obscenity, and so on; and the intrusion into life or personal liberty under Article 21, which forms the *bedrock of the privacy guarantee*, would have to be just, fair, and reasonable.

[10] IndraStra Global, 'An Analysis of Puttaswamy: The Supreme Court's Privacy Verdict', 18 November 2017, available at https://medium.com/indrastra/an-analysis-of-puttaswamy-the-supreme-courts-privacy-verdict-53d97d0b3fc6, last accessed on 18 February 2020.

Lastly, the court mentioned a fourth test for deciding whether privacy was breached—the *highest standard of scrutiny which* can be justified only in case of a *compelling state interest*.

The government, on its part, constituted a committee of experts to deliberate on a data protection framework for India in July 2017 under the chairmanship of Justice B.N. Srikrishna, former judge of the Supreme Court. The committee's mandate was to develop a framework to 'ensure the growth of the digital economy while keeping personal data of citizens secure and protected'. The committee submitted its report in July 2018 along with a draft Personal Data Protection Bill, 2018. Table 5.1 presents the key recommendations of the Srikrishna report.

Post this report, the union government prepared a Draft Personal Data Protection Bill in 2018 and the ministry solicited public feedback. The bill was finally introduced in the Lok Sabha on 11 December 2019. It seeks to provide for the protection of personal data of individuals and establishes a Data Protection Authority for the purpose. The bill is applicable to the processing of personal data by the government and Indian and foreign companies. It lays down the obligations of data fiduciaries and social media intermediaries and includes penalties for breaking the law. Box 5.2 lays down the key features of the bill.

Table 5.1 Key Recommendations of the Justice Srikrishna Committee Report

Obligations of fiduciaries (who has access to the data)	To prevent abuse of power by service providers or fiduciaries, the law should establish their basic obligations, including (i) the obligation to process data fairly and reasonably, and (ii) the obligation to give notice to the individual at the time of collecting data to various points in the interim.
Definition of personal data	It defined personal data to include data from which an individual may be identified or identifiable, either directly or indirectly and also made a distinction between personal data protection and the protection of sensitive personal data, since its processing could result in greater harm to the individual. Sensitive data means matters which require higher level of privacy (for example, caste, religion, and sexual orientation of the individual).
Consent-based processing	Consent must be treated as a pre-condition for processing personal data. Such consent should be informed or meaningful. Further, for certain vulnerable groups, such as children,

(Cont'd)

	and for sensitive personal data, a data protection law must sufficiently protect their interests, while considering their vulnerability, and exposure to risks online. Further, sensitive personal information should require explicit consent of the individual.
Non-consensual processing	Where it is not possible to obtain consent of the individual, separate grounds may be established for processing data without consent. Four bases for non-consensual processing were identified: (i) where processing is relevant for the state to discharge its welfare functions; (ii) to comply with the law or with court orders in India; (iii) when necessitated by the requirement to act promptly (to save a life, for instance); and (iv) in employment contracts, in limited situations (for example, where giving the consent requires an unreasonable effort for the employer).
Participation rights	Based on the principles of autonomy, self-determination, transparency, and accountability, the individual rights have been categorized in the following manner: (i) the right to access, confirm, and correct data, (ii) the right to object

	to data processing, automated decision-making, direct marketing and the right to data portability, and (iii) the right to be forgotten.
Enforcement models	Establish a regulator to enforce the laws who will have the power to inquire into any violations of the data protection regime, and take action against any data fiduciary responsible for the same. The regulator may also categorize certain fiduciaries as significant data fiduciaries based on their ability to cause greater harm to individuals. Such fiduciaries will be required to undertake additional obligations.
Amendments to other laws	Relevant allied laws such as the Information Technology Act, 2000, and the Census Act, 1948, which process personal data, need to adhere to the minimum data protection standards mentioned in the bill. In the event of inconsistency, the standards set in the data privacy law will apply to the processing of data. It also recommended amendments to the Aadhaar Act, 2016, to bolster its data protection framework.

Source: Justice Srikrishna Committee Report; authors' own analysis.

Box 5.2 Key Features of the Personal Data
Protection Bill, 2019

- **Applicability:** The bill seeks to provide for protection
  of personal data of individuals, and is applicable to the
  processing of personal data by (i) the government,
  (ii) companies incorporated in India, and (iii) foreign
  companies dealing with personal data of individuals
  in India.

- **Definitions:** Personal data is data which pertains
  to characteristics, traits or attributes of identity,
  which can be used to identify an individual. The bill
  categorizes certain personal data as sensitive personal
  data. This includes financial data, health data, genetic
  data, biometric data, caste, religious or political
  beliefs, or any other category of data specified by the
  government, in consultation with the authority and
  the concerned sectoral regulator.

- **Obligations of data fiduciary:** A data fiduciary is
  an entity or individual who decides the means and
  purpose of processing personal data. Such processing
  will be subject to certain purpose, collection,
  and storage limitations. Additionally, all data
  fiduciaries must undertake certain transparency and
  accountability measures such as (i) implementing
  security safeguards (such as data encryption and
  preventing misuse of data), and (ii) instituting

grievance redressal mechanisms to address complaints of individuals. They must also institute mechanisms for age verification and parental consent when processing sensitive personal data of children.

- **Rights of the individual:** The bill sets out certain rights of the individual (or data principal). These include the right to (i) obtain confirmation from the fiduciary on whether their personal data has been processed; (ii) seek correction of inaccurate, incomplete, or out-of-date personal data; (iii) have personal data transferred to any other data fiduciary in certain circumstances; and (iv) restrict continuing disclosure of their personal data by a fiduciary, if it is no longer necessary or consent is withdrawn.

- **Grounds for processing personal data:** The bill allows processing of data by fiduciaries only if consent is provided by the individual. However, in certain circumstances, personal data can be processed without consent. These include the following: (i) if required by the state for providing benefits to the individual, (ii) legal proceedings, (iii) to respond to a medical emergency.

- **Social media intermediaries:** The bill defines these to include intermediaries which enable online interaction between users and allow for sharing of information. All such intermediaries which have

(*Cont'd*)

users above a notified threshold, and whose actions can impact electoral democracy or public order, have certain obligations, which include providing a voluntary user verification mechanism for users in India.

- **Data protection authority:** The bill sets up a data protection authority which may (i) take steps to protect interests of individuals, (ii) prevent misuse of personal data, and (iii) ensure compliance with the bill.

- **Transfer of data outside India:** Sensitive personal data may be transferred outside India for processing if explicitly consented to by the individual, and subject to certain additional conditions. However, such sensitive personal data should continue to be stored in India. Certain personal data notified as critical personal data by the government can only be processed in India.

- **Exemptions:** The central government can exempt any of its agencies from the provisions of the Act: (i) in interest of security of the state, public order, and sovereignty and integrity of India and friendly relations with foreign states, and (ii) for preventing incitement to commission of any cognizable offence (that is, arrest without warrant) relating to the preceding matters. Processing of personal data is also exempted from provisions of the bill for certain other purposes such as (i) prevention, investigation, or prosecution of any offence, or (ii) personal, domestic, or (iii) journalistic purposes. However, such processing must be for a specific, clear, and lawful purpose, with certain security safeguards.

- **Offences:** Offences under the bill include: (i) processing or transferring personal data in violation of the bill, punishable with a fine of INR 15 crore or 4 per cent of the annual turnover of the fiduciary, whichever is higher, and (ii) failure to conduct a data audit, punishable with a fine of INR 5 crore or 2 per cent of the annual turnover of the fiduciary, whichever is higher. Re-identification and processing of de-identified personal data without consent is punishable with imprisonment of up to three years, or fine, or both.

Source: Personal Data Protection Bill, 2019; Bill Summary, PRS Legislative Research

The other related area of the digital ecosystem that has received attention in India is 'data localization'. Over the past year, the government has drafted and introduced multiple policy instruments that dictate that certain types of data must be stored in servers located physically within the territory of India. Presently, India has four sectoral policies that deal with localization requirements based on type of data for sectors including banking, telecom, and health: the RBI Notification on 'Storage of Payment System Data'; the FDI Policy 2017; the Unified Access License; the Companies Act, 2013, and its rules; the IRDAI (Outsourcing of Activities by Indian Insurers) Regulations, 2017; and the National M2M Roadmap.

At the same time, 2017 and 2018 have seen three separate proposals for comprehensive and sectoral localization requirements based on the type of data across sectors including the draft Personal Data Protection Bill 2018, draft e-commerce policy, and the draft e-pharmacy regulations. The policies reflect objectives such as enabling innovation, improving cyber security and privacy, enhancing national security, and protecting against foreign surveillance.[11]

While there is merit in data localization for reasons of preserving data sovereignty, there are risks that should

---

[11] Arindrajit Basu, Elonnai Hickok, and Aditya Singh Chawla, 'The Localisation Gambit: Unpacking Policy Measures for Sovereign Control of Data in India', Centre for Internet and Society, 19 March 2019, available at https://cis-india.org/internet-governance/resources/the-localisation-gambit.pdf, last accessed on 18 February 2020.

be considered. These include impact on India's trade relationship, security risks (storing data in multiple physical centres increases the physical exposure to exploitation by individuals physically obtaining data or accessing the data remotely), and economic fallout (it would increase entry barriers and compliance cost for foreign service providers).

India's AI regulations are still at a nascent stage and need to evolve significantly on issues related to human–AI collaboration, general liability frameworks, fairness appraisals, explicability standards, and safety considerations. The government needs to collaborate with relevant stakeholders, especially AI practitioners, to evolve standards and guidelines to ensure that AI technology remains socially beneficial while contributing to the economic growth of the country.

What Are the Trends in AI Regulation Globally?

AI regulation around the world today is characterized by soft approaches either aimed at incentivizing innovation in the manufacturing or digital sectors or encouraging breakthrough research. The ethical implications of AI are either regulated through specific AI codes in companies concerned with good corporate social responsibility, in research institutes (private or public) concerned with ethical research and innovation, or not regulated at all.

However, data privacy has received attention in most countries.

The USA's approach to AI regulation has been incremental, and has focused around specific pieces of legislation related to particular technologies such as drones and automatic vehicles. It has introduced legislations such as Future of AI Act, 2017, and AI Jobs Act, 2018. It does not have an overarching data protection framework. But the courts have collectively recognized a right to privacy by piecing together the limited privacy protections reflected in the first, fourth, fifth, and fourteenth amendments to the US constitution. With regard to the private sector, while no omnibus legislation exists, it has sector-specific laws that have carefully tailored rules for specific types of personal data.

The EU has a number of regulations related to financial markets (EU Markets in Financial Instrument Directive), unmanned aircraft systems (drones), motor insurance third-party liability, and liability for defective products in light of automatic vehicle technology.

The EU's General Data Privacy Regulation (GDPR) came into force from 25 May 2018. This replaces the Data Protection Directive of 1995. It is a comprehensive legal framework that deals with all kinds of processing of personal data while delineating rights and obligations of parties in detail. It is both technology and sector-agnostic and lays down the fundamental norms to protect the privacy of Europeans, in all its facets. The EU applies its jurisdiction to any personal data processing, in the EU or abroad, that originates in the EU. The GDPR also establishes penalty rates for non-compliance, rules on user consent, data erasure, breach notification, right to access, and data portability. But importantly, the GDPR allows for the flow of data to third-party countries *if* the receiving country's laws are in compliance with the GDPR's rules.

China's AI regulation is not yet in the public domain, but it has primarily approached the issue of data protection from the perspective of averting national security risks. Its cybersecurity law, which came into effect in 2017, contains top-level principles for handling personal data. A follow-up standard (akin to a regulation) issued earlier this year adopts a consent-based framework with strict controls on cross-border sharing of personal data. It remains to be seen how such a standard will be implemented.

In a Nutshell

The promise of AI will be fulfilled if it is able to address a large number of society's most pressing problems—

poverty, climate change, and access to health care and education to name a few. In order for this to happen research needs to focus on these areas rather than only on making AI technology more capable.

The promise of AI can also turn into disaster if predictions about job losses are not taken seriously by countries across the globe. Although some reports do predict that more jobs will be created than lost, the challenge is to ensure that those jobs are open to all. Therefore, countries need to find ways to retrain and reskill large sections of the population so that they are AI ready.

As both the promise and risks of AI are likely to have their impact across countries, industries and social classes, governments need to be proactive in not only harnessing AI technology for economic growth but also in putting in place regulations to ensure that citizens are protected from the threats posed by AI. However, given the early stage of AI development, it is important to focus on laws and norms that retain flexibility as new possibilities and problems emerge. This is particularly crucial given that AI is multipurpose in nature. It is also imperative that countries cooperate and collaborate with each other at various levels—government, academia, civil society, and corporates—to develop regulatory frameworks that address the challenges and risks posed by AI. The spillover effects of contradictory regulations across countries and working in silos could be immense given the scale of threats that AI can pose on the security and sovereignty of a country.

Every regulation that is developed needs to debate the trade-offs between many factors: how stringent should standards of explainability be? What should be the definition of fairness as there are conflicting definitions? How should safety problems be addressed? But in setting benchmarks, it is important to factor in the opportunity cost of not using an AI solution when one is available; and to determine at what levels of relative safety performance AI solutions should be used to supplement or replace existing human ones. AI systems can make mistakes, but so do people, and in some contexts, AI may be safer than alternatives without AI, even if it is not fail-proof.

Finally, we need to keep in mind that AI is a tool that can be applied with good or ill intent. Therefore, it is important to think of the ethical implications of AI while designing it. Similarly, we need to find a balance between regulations that protect citizens while also not impeding technological breakthroughs.

Bibliography

Bostrom, N. 2014. *Superintelligence: Paths, Dangers, Strategies*. Oxford: Oxford University Press.

Brynjolfsson, E., and McAfee, A. 2018. *The Second Machine Age: Work, Progress, and Prosperity in a Time of Brilliant Technologies*. Vancouver B.C.: Langara College.

Harari, Y. N. 2018. *Homo Deus: A Brief History of Tomorrow*. New York: Harper Perennial.

Hofstadter, D. R. 2006. *Gödel, Escher, Bach: An Eternal Golden Braid*. New York: Basic Books.

Kurzweil, R. 2016. *The Singularity Is Near: When Humans Transcend Biology*. London: Duckworth.

Lee, Kai-Fu. 2019. *AI Superpowers: China, Silicon Valley, and the New World Order*. S.l.: Mariner Books.

Mcafee, Andrew. 2018. *Machine, Platform, Crowd: Harnessing Our Digital Future*. W.W. Norton.

Reese, B. 2018. *The Fourth Age: Smart Robots, Conscious Computers, and the Future of Humanity*. New York: Atria Books.

Tegmark, M. 2018. *Life 3.0: Being Human in the Age of Artificial Intelligence*. London: Penguin Books.

Wilson, H. J., & Daugherty, P. R. 2018. *Human + Machine: Reimagining Work in the Age of AI*. Boston: Massachusetts Harvard Business Review Press.

Index

About the Authors

Kaushiki Sanyal is co-founder and CEO at Sunay Policy Advisory Pvt. Ltd, a Gurgaon-based public policy consulting, training, and research start-up. She has previously worked for NDTV, Capital IQ, PRS Legislative Research, and the Indian School of Business. Educated at Lady Sriram College and Jawaharlal Nehru University, New Delhi, where she earned her PhD, Kaushiki has co-authored two books: *Oxford India Short Introductions: Public Policy in India* (2016) and *Shaping Policy in India: Alliance, Advocacy, Activism* (2017). She has also contributed several chapters for books and journal articles.

Rajesh Chakrabarti is dean at the Jindal Global Business School, O.P. Jindal Global University, India, and co-founder at Sunay Policy Advisory Pvt. Ltd. He has been a member of the faculty at Georgia Tech,

USA, University of Alberta, Canada, and the Indian School of Business. Rajesh is an alumnus of Presidency College, Kolkata, and IIM Ahmedabad and earned his PhD from the University of California at Los Angeles. He has either authored or edited nine earlier volumes and published several articles in top international research journals in management and economics.